Astronomers' Universe

Series Editor
Martin Beech, Campion College, The University of Regina
Regina, SK, Canada

The Astronomers' Universe series attracts scientifically curious readers with a passion for astronomy and its related fields. In this series, you will venture beyond the basics to gain a deeper understanding of the cosmos—all from the comfort of your chair.

Our books cover any and all topics related to the scientific study of the Universe and our place in it, exploring discoveries and theories in areas ranging from cosmology and astrophysics to planetary science and astrobiology.

This series bridges the gap between very basic popular science books and higher-level textbooks, providing rigorous, yet digestible forays for the intrepid lay reader. It goes beyond a beginner's level, introducing you to more complex concepts that will expand your knowledge of the cosmos. The books are written in a didactic and descriptive style, including basic mathematics where necessary.

More information about this series at http://www.springer.com/series/6960

John Etienne Beckman

Multimessenger Astronomy

 Springer

John Etienne Beckman
Instituto de Astrofísica de Canarias
San Cristóbal de la Laguna, Sta. Cruz de Tenerife, Spain

ISSN 1614-659X ISSN 2197-6651 (electronic)
Astronomers' Universe
ISBN 978-3-030-68371-9 ISBN 978-3-030-68372-6 (eBook)
https://doi.org/10.1007/978-3-030-68372-6

Cover illustration: The remnant of Supernova 1987a in radiation at very different wavelengths: submillimetre wave data from ALMA (coded in red) shows newly formed dust in the centre of the remnant. Hubble Space Telescope visible image (coded in green) and Chandra Space Telescope X-ray image (coded in blue) show the expanding shock wave. Credits: HST and Chandra images: NASA, ALMA image: ESO/NSF/NINS.
Superposed graph: The detection of neutrinos from Supernova 1987a by the Kamiokande neutrino detector. Horizontally dispersed points show the background neutrino level, and the vertical pulse in the centre shows the clear detection of the neutrinos from the supernova. Credit: Kamioka Observatory, ICRR (Institute for Cosmic Ray Research), The University of Tokyo Design: Inés Bonet/IAC/UC3

This Springer imprint is published by the registered company Springer Nature Switzerland AG.
The registered company address is: Gewerbestrasse 11, 6330 Cham, Switzerland

Dedicated to Leocadia Pérez González and Jaime Beckman Pérez.

Foreword

John and I first met at the National Scientific Ballooning Facility in Palestine Texas. We were both trying to fly experiments that would measure the spectrum of the cosmic microwave background that had been discovered a few years before. The critical thing we both were anxious to establish was whether the spectrum was truly that of a black body at around 3 degrees Kelvin. In those days, the measurement was really difficult; the technology for observations at a wavelength around a millimetre was not well developed. You had to make your own instruments from scratch, and furthermore, the detectors as well as the instruments had to operate at cryogenic temperatures, comparable if not smaller than the temperature of the background radiation itself. The most serious enemy was the radiation from everything at room temperature (300 degrees Kelvin). Even a little hole or a sidelobe of the beam touching a warm surface was enough to wreck the measurement. You could not observe from the ground or even a high mountain so we both chose to fly our instruments on balloons that made it to an altitude of 40 km and gave typically 6 hours of night-time observing at the edge of the atmosphere.

John and his group had made an elegant polarising interferometer based on a design by his colleague Derek Martin while my group used a simple set of bandpass filters; we both used the same kind of detector. John's first effort did not get any data due to a destructive balloon failure while our flight functioned but gave a spurious result due to high frequency leakage in the critical channel designed to measure the thermal peak of the 3 K background. Both of us continued with further flights. In a subsequent flight, John saw a peak in the CMB spectrum but was unsure of the calibration. In our case, we needed to subtract a significant amount of atmospheric radiation to show the peak. After several years of flights by many more groups and instruments with slowly improving results, the ground work had been set to convince NASA to actually carry out a dedicated space mission to measure the spectrum and angular distribution of the cosmic background as well as an assay of the sky in the infrared. This was the COBE mission first proposed in 1972 by John Mather but finally flown in 1989. The instrument used to measure the spectrum on that mission was derived from the interferometer John had flown.

John and I met again a few years ago when he invited me to come to the Almeria Astronomy Week held in Almeria, Spain, in June 2018 as part of a celebration of the first direct detection of gravitational waves. We gave a talk together to a group of amateur astronomers who were primarily Spanish and more comfortable with a talk in Spanish rather than English. I was a little wary of the idea, but it worked well. I would show a slide and describe its content in English for a few minutes after which John would translate with the proper emphasis and equivalent excitement including the ironic comments and asides. What impressed me most was his depth of understanding and ability to explain. It was during that visit John told me about his ambitious plan to write this book.

The reason I am telling you all of this is that what John learned through that experience with the 3 K background coupled with his breadth and knowledge is all over this book and is what makes it a unique book about astronomy. (If you are interested in his interferometer you will find it in the section on infrared astronomy where he used it on the Concorde supersonic aircraft to look at the solar chromosphere.) John tells you about the astronomy, but he also explains the physics and other sciences associated with the sources. You will also learn about the instruments used to make the observations and enough about the physics and engineering to appreciate how they work and some of the human ingenuity involved in their design.

I learned a lot from the book, especially about some of the major puzzles in current astronomy and astrophysics, enough so that you are well enough equipped, should you want, to launch into some of the technical papers about the topics in the literature. Above all John elegantly shows the extent of the science involved in understanding our universe and its contents. In the next edition of the book, there will probably be a chapter on biology and the possible evidence for life in many places in the universe. Enjoy the read.

Newton, Massachusetts, USA   Rainer Weiss
November 2020   Emeritus Professor of Physics, MIT
                                        Nobel Laureate in Physics, 2017

Preface

Astronomy has the power to inspire science, art, religion, and popular culture. It is one of the few sciences where amateurs can and continually do make significant contributions, and it is the originator of citizen science, in which the general public can contribute to discoveries via computer-based techniques. It is one of the oldest and most traditional of the sciences but at the same time is continually pushing back the frontiers of new technology, due to the technical prowess required to detect the incredibly faint signals which reach us from around the universe. From when Galileo used the first modern astronomical instrument, the telescope, until almost half way through the twentieth century all our astronomical knowledge was acquired using visible light, even though the visible wavelength range occupies only a small fraction of the full gamut of the electromagnetic spectrum. During the nineteenth century, two powerful tools were added to the astronomers' kit: spectroscopy, which led to the exploration of the composition and the physical states of objects far from the Earth, and photography, which enabled astronomers to accumulate the photons from astronomical sources, allowing them to obtain images of objects much fainter than is ever possible with the naked eye, even through a telescope, and hence to penetrate more and more deeply into space. Within the first three decades of the twentieth century, astronomers, using only visible light, already knew that there are galaxies outside the Milky Way and that on the largest scales the universe is expanding.

Radioastronomy was the first of what came to be called the "new astronomies". In the United States in the 1930s, radio experts who were also amateur astronomers had detected radio waves from outside the Earth, and can be considered the founders of radioastronomy, but this science was really lifted into reality by the technical advances in radio made during the Second World

War to create radar. I was a boy in the 1940s and 1950s and interested in astronomy from an early age. Living in England in those years I learned about this new window on the universe, and visited the Jodrell Bank radio-observatory, which was not very far from Sheffield where I lived. The new generation of radioastronomers, together with a bright group of theorists, was setting the pace and regarded the optical astronomers as "old hat". In many ways they were right, and their influence was the key to the subsequent resurgence of British optical astronomy, when important optical telescopes had been built on sites well away from cloudy British skies. In the meantime, radioastronomers in the Netherlands, the United States, Australia, and the United Kingdom were mapping large parts of the Milky Way for the first time in the radio emission from atomic hydrogen, penetrating not only the cloudy skies but also the dust which shrouds many of the most interesting astronomical sources in the Galactic plane. They were also making a series of major sky catalogues in radio. These included the famous series by the Cambridge group under Ryle, in a series which reached 7C but whose most famous 3C catalogue included many sources which did not correspond to anything known in the visible. In the early days of radioastronomy, in the early 1950s, these were often referred to as "radio stars", but as time went on, and other techniques came into use, almost all of them were found to be quite different from stars.

For two decades, radioastronomy and optical astronomy were the main agents of advance in the field. During this period, two striking discoveries were made by the radioastronomers: quasars and pulsars. Quasars were "quasistars" bright point sources of radio energy which did not correspond in position to any known stars, even faint stars. At radio wavelengths, it was difficult to pinpoint the position of any source, because the angular resolution of single radiotelescope was nowhere near good enough. A breakthrough came when, in 1962, one of the strong sources 3C273 (number 273 in the third Cambridge calalogue) could be observed when occulted by the Moon, and by careful timing of the moment when the source disappeared and reappeared its position could be measured with only a small area of uncertainty on the sky. The optical observers could then identify this source with a quite faint starlike object, but whose velocity of recession from us placed it at a redshift of 0.16; it was apparently receding from us at 16% of the velocity of light. This was an astonishing result, because from this value of redshift using the Hubble–Lemaître law of the expanding universe, its distance was over 2000 million light years, and the power it was emitting had to be similar to that of a galaxy rather than of a star. We are now accustomed to thinking of these objects as the supermassive black holes which are at the centres of galaxies, and which produce colossal energies in a compact volume due to the conversion of gravitational energy into other types of energy. But this was one of the signs that opening a new window on the universe, in this case the radio window, would lead to important discoveries. Five years later, another new type of astronomical objects, the pulsars, was discovered by the radioastronomers, and these rapidly spinning, highly compact stars turned out to be neutron stars, the central cores of much more massive stars which have exploded as supernovae. Pulsars have played a special role in showing that general relativity is, so far, the best description of gravity, and they remained the best demonstrations of general relativity in action until gravitational waves were first measured in 2015.

After radioastronomy the electromagnetic windows on the universe began to open one after another. As the Earth's atmosphere is fully or partly opaque to radiation in all the wavelength ranges except the narrow optical waveband, and a large fraction of the wider radio waveband, many of the advances were made when it became possible to escape from the effects of the atmosphere (it is, of course, no coincidence that the atmosphere is transparent in the optical, i.e. the visible, wavelength range; evolution has ensured that animals' eyes are most sensitive in this range). This was done by using balloons, aircraft, rockets, and above all satellites as observatories. The era of astronomy from space beginning in the second half of the twentieth century is by any standards the golden age of astronomy. The new opportunities stimulated the invention of new techniques, covering notably the infrared, X-rays, and gamma-rays. The visible and the ultraviolet also benefitted from the development of electronic detectors which substituted the photographic plate, yielding orders of magnitude in improved sensitivity. By the turn of the century, astronomy and astrophysics were technically capable of tackling some of the most interesting problems in the whole of physics, combining information about the universe, its initial phases, and its evolution, with the puzzles intriguing particle physicists. The cosmic microwave background radiation, discovered in 1964 at short radio wavelengths, had been shown to be the most directly observable relic of the Big Bang, and its study, combined with optical and infrared studies of stars, galaxies, and the intergalactic medium on the largest scales, put cosmology on a scientific basis, as opposed to the largely speculative models relying on good theory (general relativity) but with limited types of observations that had prevailed for the previous half century. Even so, all the information gathered from the cosmos was via electromagnetic waves, albeit over a range of more than 20 orders of magnitude in wavelength (also in frequency, and in energy). In one sense, we were still relying almost entirely on a single type of messenger to deliver the information about the universe. Much of this book is aimed at describing the practical ways in which these new astronomies worked.

There was in fact one other stream of information from outside the Earth which had been actively observed since the first decade of the twentieth century. These were the cosmic rays, which was a misnomer for the high energy particles which impinge on the Earth from space. These particles were first detected using relatively simple equipment which had been developed during the beginnings of the new science of radioactivity, which would become nuclear physics. These particles have energies ranging well above those which can be produced even in the biggest particle accelerators, and until the mid-1950s they were used to make new discoveries in particle physics rather than to explore astronomical objects. They were much cheaper to use than to build accelerators, hence of great interest to particle physicists, but their sources could not be easily identified, hence of less direct interest to astrophysicists. Particles such as protons (hydrogen nuclei), electrons, and alpha particles (helium nuclei) have their trajectories interfered with by magnetic fields as they pass through interstellar and intergalactic space, so it is almost impossible to discover their sources directly. These particles make up most of the cosmic rays, and as we will see, they originate in some of the most energetic processes in the universe, giving us unique insights into those processes. They also make up a large part of what is found as radioactivity on Earth, and their names, alpha particles, beta particles, (which are electrons), and gamma-rays reflect experimental discovery and classification in the late nineteenth century prior to the understanding of their nature and detailed properties. We now know that gamma-rays are the highest frequency, highest energy form of electromagnetic radiation. This means that they travel through space just as light does, with no deviation by magnetic fields, for example. We also know that they are produced in similar processes to those which produce cosmic rays, and indeed in one sense they are the only true cosmic rays, while the other types should really be called cosmic particles. So we can combine detections of gamma-rays and cosmic rays to find out about high energy processes, such as supernova explosions, and accretion onto black holes, because the gamma-rays allow us to detect the position of the source in the sky.

Among the most elusive sub-atomic particles are the neutrinos. They were first predicted to exist in the 1930s in order to explain the missing energy and momentum in certain nuclear reactions, but their interaction with matter in general is so weak that it was very difficult to find them. They were not detected until the 1950s as products of a nuclear reactor which was producing neutrinos in such huge quantities that a few of them interacted with the apparatus of the physicists who were looking for them. It was during the same period that nuclear physics was able to give a detailed explanation of the processes which produce the energy of the stars. For stable stars such as the Sun the process is the fusion of four hydrogen nuclei to produce one helium nucleus. This releases energy in the Sun's core in the form of gamma-rays and also neutrinos; two neutrinos are produced for every helium nucleus formed. These escape easily from the Sun and travel towards Earth at almost the speed of light. The number of fusions per second in the centre of the Sun is close to 10^{38} which means that the number of neutrinos passing through a square metre at the Earth is some is 10^{15} per second. It is fortunate for us that neutrinos interact so weakly with all other particles because this number is close to the number of neutrinos which pass through your body every second. The first step in neutrino astronomy was taken in the 1960s when a small number of solar neutrinos were detected with a "neutrino telescope" down a deep mine. Since then, the study of solar neutrinos has told us a lot about neutrinos, and something about the Sun's interior. In this way, another type of messenger has been established. High energy neutrinos from supernova explosions have also been detected, and neutrino telescopes are now present deep in the Mediterranean and under the Antarctic ice sheet.

The event which brought the phrase "multimessenger astronomy" to the attention of the media was the detection of gravitational waves in 2015. In the nineteenth century, Maxwell had formulated the theory of electromagnetism which associated the movement of electric charge (later shown to consist of electrons) with waves of energy transmitted through space by radiation. Oscillating electrons produce electromagnetic waves, and it is these waves which astronomers use, in their wide variety of forms from radio waves to gamma-rays, to study the phenomena in the universe. By analogy Einstein's theory of gravitation associates mass in motion with the emission of gravitational waves. But the gravitational force is very weak compared to the electromagnetic force, and to produce detectable gravitational waves requires huge masses in rapid motion. Einstein himself, after first doubting his own prediction, eventually came to believe that gravitational waves should be a real phenomenon, but too tiny compared to electromagnetic and nuclear phenomena to be ever detectable. But in the decades which have passed since his death decisive steps have been made to show that gravitational waves are a reality and not a mere theoretical construct. Some of these steps are observational: the detection by conventional techniques of very compact objects with large gravitational fields such a neutron stars and black holes. The other step is technical, the construction of detectors sensitive enough to respond to the tiny signals emitted when such compact objects merge, at which event they emit a large pulse of gravitational waves. I think that the technical achievement which enabled gravitational waves to be detected is fully up to the level of the intellectual achievement which produced general relativity. Once the first detection was made, gravitational wave astronomy has taken its place as a powerful exploration method to deepen our understanding of the zones in the universe where extreme gravitational fields produce extremely energetic phenomena.

But in the general and fully justified excitement about the new ways of probing the universe embodied in neutrinos and gravitational waves, most people have forgotten another way we have to examine the universe, and to do so directly in the laboratory. This is the use of meteoritic material. Meteorites are mostly the dusty and rocky remains of comets which have been pushed into interplanetary space from the surface of a comet when it heats up as it approaches the Sun. These remains are left in the comets' orbits, and if the Earth's annual path around the Sun cuts one of these orbits, the particles fall through the atmosphere and the larger ones land on the ground. The light trails caused by the burning up of the particles are meteors, and the rocks which reach the ground are meteorites. In the past fifty years, nuclear chemistry has been applied to meteorites to derive their ages, and conventional chemistry has analysed their compositions. The information gained tells us a great deal about the history of the solar system, including material which formed the disc of particles which gave rise to the planets, and even particles whose ages predate that of the Sun. More recently space missions have allowed scientists to collect similar samples directly from the Moon, from comets, and from asteroids. Meteorites have been found on Earth which are clearly of Martian origin, and NASA's Mars rover series has enabled *in situ* analysis of Martian rock and soil. I have put all of these techniques into one basket, labelled "hands-on astronomy", and this is surely a very effective messenger from outside the Earth, and from the past.

When I was considering writing this book I wanted to give a wide meaning to the term "multimessenger" which meant including not only the methods of astronomy which do not use electromagnetic radiation, but also the full range of those that do. The initial idea was to place emphasis on the techniques, on how the measurements were made, and the second idea was to present some brief historical lines of development. But I quickly realised that a complete book based on these two principles would be encyclopaedic, and well beyond my scope. So I had to find criteria for choosing only a small fraction of what could have been included. I am afraid to say that I adopted the criterion of including themes and details that I find of personal interest. This means leaving out large swathes of information which are very interesting too. This way of choosing also meant that I have done less than justice to those areas where my own knowledge is least. Even though I have an eclectic interest in astronomy, it would be absurd to claim that I have more than a thin veneer of knowledge in certain areas. This is particularly the case for three of the chapters in the book, on ultraviolet (UV) astronomy, on gamma-ray astronomy, and on "hands-on" astronomy. I have based two of these chapters on specific review articles, whose authors I have recognised in the acknowledgements. For the third, I was fortunate enough to find an expert willing to give me detailed help, which I have also acknowledged.

It would not have made sense to write only about the techniques and their development without including some of the discoveries. As I went along writing each chapter and finding illustrations of the objects observed, I realised that many of the objects brought into focus by the new messengers are not "normal" stars or planets, and many of them are not even galaxies. The tendency is to highlight very energetic objects which show up especially in the interstellar medium. These can be supernova remnants, the zones where supernovae have exploded in the recent past, leaving luminous gas expanding at hundreds, even thousands of kilometres per second. This gas can be detected in the optical range, by the emission lines from its ionised elements, but it also emits strong radio waves, due to a process called synchrotron emission, involving magnetic fields, also X-rays and gamma-rays, and the dust made by the heavy elements created in the supernova emits in the infrared. The Milky Way has considerable numbers of these supernova remnants, but also regions of active star formation which glow over a wide wavelength range, notably over the full infrared and submillimetre ranges, also due to the interstellar dust produced in the regions. The Milky Way is in many ways the star performer here, matched in number and variety of images by the Sun, which over the years has served as a test-bed object for high energy imaging in astrophysics. It is my hope that some of the objects imaged in the different chapters of the book will be useful as links to show unity in variety. Not all of astronomy lends itself to beautiful photographs, of the sort regularly presented in NASA's Astronomy Picture of the Day (APOD) and I have not held back in the use of graphs, some fairly technical, to explain the meaning of results, notably in the chapter on cosmology, but in general throughout the book. However, I have not included any dreaded equations and relatively few symbols, in the hope that interested non-science readers will be able to maintain their interest.

During my scientific lifetime I have been an active witness to the opening up of many of the new pathways through the universe described here. This book can be an initial guide, and for those readers who are young and want to take up the challenge, I hope that they will be able to venture along some of the pathways, intellectually or even physically, and discover the immense territories awaiting humanity in space.

San Cristóbal de la Laguna,   John Etienne Beckman
Sta. Cruz de Tenerife
December 2020

Acknowledgements

I have used many sources of information and images while preparing this book, and the respective attributions are given in the figure captions. The figures which have been taken from the major journals: *The Astrophysical Journal*, *Astronomy & Astrophysics*, and *Monthly Notices of the Royal Astronomical Society* all are used with explicit permission from those journals and their governing entities, as well as their authors. But some of the help merits special thanks to individuals. Firstly, I would like to thank the LIGO collaboration for permission to use figures from their work, Martin Hendry for helping me to find good quality figures from their archive, and above all Rainer Weiss for helping me with both the historical and the technical content.

For the chapter dealing with the analysis of meteoritic material I am very pleased to acknowledge the help of Larry Nittler, who supplemented my quite thin previous knowledge with the benefit of his deep and wide experience, as well as providing me with figures. I have to thank the Antares, Ice Cube, Kamiokande, SNOLab, the AMS experiment, the MAGIC telescopes, the CTA observatories, and the Pierre Auger observatory for permission to use images. I have turned to two review articles for the chapters where my previous knowledge needed a boost, and I am grateful to the authors of these for generously allowing me to learn from, and use, their material. The articles are: for chapter 4 on UV astronomy, the review by Jeffrey Linsky "*UV astronomy throughout the ages: a historical perspective*" published in *Astronomy and Space Science*, Volume 363, page 101, in 2018, and for chapter 6 on gamma-ray astronomy, the review by Stefan Funk, "*Ground- and Space-based Gamma-Ray Astronomy*" published in *Annual Reviews of Nuclear and Particle Science*, Volume 65, page 245, 2015. In both cases, there is much more in the reviews than I have been able to select for the chapter in this book, so they are recommended reading for those interested.

I benefitted from the help in drawing figures by Inés Bonet, of the Unit for Science Culture and Outreach at the Instituto de Astrofísica de Canarias (IAC). My friend and colleague Terry Mahoney, also of the IAC, gave me much useful guidance from his wise publishing experience, and helped me considerably in stitching the book together. I would also like to thank Rebecca Roth, Image Coordinator and Social Media Specialist at NASA's Goddard Space Flight Center, for her help with all of the NASA material needed to illustrate the book. I am also grateful for permission to use the ESA Image Gallery photo archive to contribute to the illustrations. At Springer I received constant help and encouragement from the Editors, Ramon Khanna, Rebecca Sauter, and Christina Fehling, while from Straive in Chennai I was given the needed technical help in publishing by DhivyaGeno Savariraj.

My colleagues at the Instituto de Astrofísica de Canarias have provided me with an exciting atmosphere of observational and experimental astronomy in which it would be difficult not to learn. I would like to give warm thanks to my graduate students and postdocs who over the years have educated me and kept me up to the mark as astronomy has progressed at an ever increasing rate.

Finally, I have to thank Leo, my wife, for her patience in realising that although astronomy has always competed for her affections, without her constant support I would be lost, and Jaime my son, for making my life interesting when I might have lapsed into complacency.

Contents

1

Optical Astronomy

1.1 The Instruments of Optical Astronomy

1.1.1 Telescopes

Optical astronomy, or astronomy in the visible wavelength range, is the original form of the science, which is a paradigm for all observational astronomy using electromagnetic waves. Although people had been observing and interpreting the sky for millennia, the change to modern science came with the invention of the telescope. The concept of an instrument for improving our sight over long distances was first put into practical form by the German-Dutch lens maker Hans Lippershey who in 1608 requested a patent for his combination of a converging and a diverging lens in a pair of sliding concentric tubes which produced an image of distant objects, bringing them apparently much closer to the eye. A description of this instrument reached Galileo Galilei, then a professor at the University of Pisa, who quickly made his own, and went on to build telescopes of increasing power. He used them to observe the Moon, discovering its mountains, craters, and plains, Jupiter, whose four major satellites, the Galilean moons, he observed and recognised as such, Venus, whose phases he also discovered, and the Sun, where he confirmed the presence of sunspots previously noted by Christoph Scheiner. Figure 1.1 is an optical diagram of a Galilean telescope.

A design change introduced by Johannes Kepler in 1611 was to replace the concave eyepiece by a second convex lens. This widened the field and produced an image outside the telescope which could be focused onto a plane for inspection. The drawback of these telescopes was the inversion of their images,

J. E. Beckman, *Multimessenger Astronomy*, Astronomers' Universe, https://doi.org/10.1007/978-3-030-68372-6_1

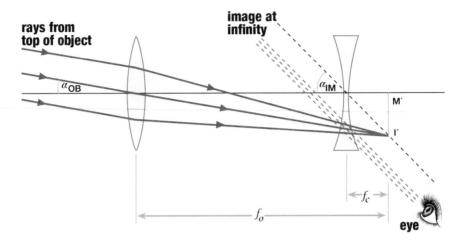

Fig. 1.1 Diagram of Galilean Telescope optics. The convex objective lens would produce an inverted image M′-I′ but the concave eyepiece lens produces a virtual image which is erect, "at infinity" and magnified compared to the object, by the ratio α_{IM}/α_{OB}. It could not be used to take photographs! Credit: The Open University/IAC-UC3

which made it less desirable for terrestrial use, but is no real problem for astronomy.

Refracting telescopes, those using lenses, played a key role in astronomy for two centuries, and many small refractors are in use around the world. But they have several drawbacks. Firstly the glass has a different refractive index for different wavelengths of light, so a simple lens brings light of different colours to different foci. This defect has been partly overcome by making "achromatic" lenses of at least two different types of glass (the same problem in cameras has been overcome in a similar way). Secondly a lens in a telescope must be supported around its edges, so a large lens will tend to deform as the telescope changes its pointing on the sky. This sets a limit to the sizes of refractors. The largest refractor ever built for professional use is at the Yerkes Observatory of the University of Chicago. It was put into operation in the early years of the twentieth century, and ceased operations in October 2018. Many of the world's greatest astronomers based their early researches on this telescope, including Edwin Hubble, Subrahmanyan Chandrasekhar, Nobel Laureate whose name was given to the major X-ray satellite Chandra, the Dutch-American astronomer Gerard Kuiper, and the outstanding populariser of astronomy Carl Sagan. Figure 1.2 shows the Yerkes refractor in 1921.

Reflecting telescopes, or reflectors, have now replaced refractors in all professional observatories, and virtually all amateur astronomers now use them. The first reflector was designed and built by Isaac Newton. The main focusing element, the primary mirror, is concave and brings the light to a focus

Fig. 1.2 The 40 in. refracting telescope at the Yerkes Observatory, U. Chicago with staff and visitor Albert Einstein in 1921 Credit: University of Chicago Photographic Archive, [apf 6-00415], Special Collections Research Center, University of Chicago Library

on the central axis of the telescope tube. Newton placed a small flat secondary mirror inside the tube, which diverts the beam and brings the focus out at the side of the tube. The scheme of a Newtonian reflector is shown in Fig. 1.3.

This works very well for a fairly small telescope where the observer can stand beside the instrument and look into the side tube. But for larger telescopes, and above all for observing with instruments, it is much more convenient to have the optics all aligned along the central axis. The basic design for this was suggested by Laurent Cassegrain in 1672, although as with all ideas, previous scientists had similar suggestions. In its most simple form a Cassegrain reflector has a concave primary mirror which sends the light to a focus back along the axis of the telescope tube, but before it reaches this focus it is intercepted by a small convex mirror which sends the light back towards the primary. A circular hole in the primary allows the light to go through to a focus just behind it, where on a modern professional telescope an instrument is placed, although for smaller telescopes an eyepiece can be put there for direct viewing. A scheme for a Cassegrain reflector is shown in Fig. 1.4. To minimise the aberrations in the image, the primary is normally a parábola and the

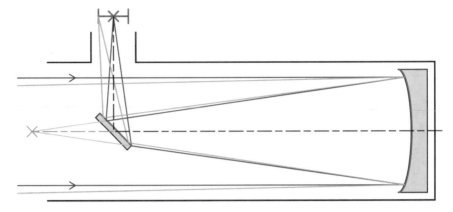

Fig. 1.3 Optical scheme of a Newtonian reflecting telescope. Light enters down the main tube, and is brought to a focus at one side, by the combined parabolic primary mirror to the right, and the small flat mirror in the centre. Credit: Newton-Teleskop. svg: Kizar derivative work: Kizar, CC BY-SA 3.0 (http://creativecommons.org/licenses/by-sa/3.0/), via Wikimedia Commons

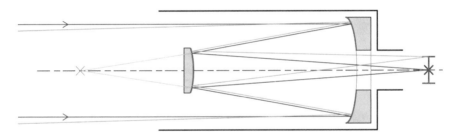

Fig. 1.4 Optical scheme of a Cassegrain reflecting telescope. Light enters from the left and first meets the primary mirror at the end of the tube. It is then focused back towards the secondary mirror, which reflects it again, to a focus behind the primary, where instruments can be placed. Credit: Upload: Wikibob, Original: ArtMechanic on de.wikipedia, CC BY-SA 3.0 (http://creativecommons.org/licenses/by-sa/3.0/), via Wikimedia Commons

secondary a hyperbola, neither of them differing greatly from simple spherical surfaces. Since the original Cassegrain design there have been a number of refinements introduced, among which the most widely used professionally is the Ritchey-Chrétien format, invented in the early twentieth century, where both primary and secondary mirrors are hyperbolic.

Reflecting telescopes were limited in their early days by the need to make the mirrors of speculum metal, an alloy of two parts copper to one part tin, which takes a reasonable reflecting surface, but tarnishes easily. Among the earliest large telescopes using speculum metal was that made in 1749 by

Fig. 1.5 The 72 in. telescope made by Lord Rosse, in Parsonstown, Ireland. He discovered the Whirlpool galaxy, M51, with this telescope. Copperplate engraving c. 1860

William Herschel, the discoverer of the planet Uranus, which had a 49.5 in. diameter primary, and a famous telescope of this type was the 72 in. telescope made by Lord Rosse, in 1845, shown in Fig. 1.5.

The process of silvering the front surface of a glass mirror, developed by von Steinheil and also by Foucault in the 1850s made reflectors more efficient, as silver reflects 90% of the incident light, compared with 65% for speculum metal. But also glass mirrors are much easier to figure, to give them the required precise shape. This was the trigger which led to the universal use of mirrors in large telescopes, although since the 1930s silver was generally replaced by aluminium, which is less subject to tarnishing in contact with the air. Even so, most major astronomical observatories re-aluminize their mirrors regularly, and have a specialised vacuum chamber on site for doing so. From the late nineteenth century to the late twentieth century single mirror reflectors of increasing size were made for telescopes, progressing from the 1.5 m ("60 in.") and 2.5 m ("100 in.") telescopes on Mount Wilson to the 5 m ("200 in.") Hale telescope on Mount Palomar in California, which was the largest telescope in the world for nearly 30 years after its inauguration in 1948. There are currently 27 single mirror reflectors working in the optical and near infrared range around the world with primary mirrors larger than 3 m in diameter, among the largest being the four telescopes of 8.2 m in diameter which make up the Very Large Telescope of the European Southern

Observatory in Chile, and Japan's 8.2 m Subaru telescope in Hawaii. But the largest optical telescopes in the world now have segmented primaries, composed of smaller mirrors combined into a single effective large mirror. This is easier to handle mechanically for the largest sizes, but requires continuous sophisticated electronic control to maintain optical perfection as the telescope pointing moves around the sky. We cannot give a list of these largest telescopes, but mention only the two US telescopes, the Kecks, on Hawaii since 1993 and 1996 respectively, each with 10 m diameter primaries made up of 36 segments, and the largest current optical/infrared telescope in the world, the 10.4 m Spanish GRANTECAN on the Canary Island of La Palma, operational since 2009, which also has a primary composed of 36 segments. Figure 1.6 shows two large optical telescopes: the equatorially mounted Hale Telescope, and the altazimuthally mounted GRANTECAN.

The advent of computer chips brought about various advances in the design of telescopes but one basic contribution is worth explaining here. Until the 1970s all telescopes, including the largest, were mounted equatorially, but the incorporation of computer control allowed the basic way of mounting telescopes to be changed to altazimuth. The difference has made it possible to build the modern super-telescopes. To point a telescope to any position on the sky and then keep it locked onto an object, it needs two axes of rotation. From the middle of the nineteenth century the best way to organise this was the equatorial mount, invented by Joseph von Fraunhofer in Germany. In this mount one of the axes, the polar axis, is parallel to the Earth's axis of rotation, and the other is perpendicular to it. To follow a star, or other celestial object, the telescope is rotated about these two axes until the object is found, then the perpendicular axis is held fixed, and the telescope is rotated about the polar axis at a steady rate to counteract the Earth's rotation. The system needs only a single driver, in modern times an electric motor, to follow the object. There are several ways to arrange the two axes; one of these is the way Fraunhofer did it, with the telescope on one arm of the perpendicular axis, counterbalanced by a weight on the other arm, which can be varied to suit the weights of the instruments placed at the telescope focus; this is best suited to small telescopes. There are half a dozen variants of equatorial mounts, each with its advantages and inconveniences. The 200 in. Telescope on Mount Palomar was one of the last generation of large telescopes using an equatorial mount, and because of its weight it used a fork mount The reason why equatorial mounts were universal in their time is that the telescope can be driven with a single motor at constant rate, which simplifies most of the design. On the contrary an altazimuth mount, where the two axes are locally vertical and horizontal, needs two motors, both driven at variable rates, to follow an object.

Incorporating computer technology into the drives solves this difficulty, and allows most other features of the telescope to be simplified. Equatorially mounted telescopes are on inclined axes, leading to awkward imbalance, and entailing a much larger telescope dome. The weight of a large telescope, tens of tons, bears down vertically on an altazimuth mount, which allows much bigger telescopes to be built. The first major altazimuth telescope was the Russian 6 m telescope in the Caucasus, operational since 1975, and all subsequent large telescopes, including the GRANTECAN featured in Fig. 1.6 have this design. Nowadays even small portable amateur telescopes are made altazimuth, as this allows them to be transported and used at different latitudes. At the other end of the size range the astronomical communities are now well advanced in the process leading to the construction of telescopes of new giant dimensions: the Extremely Large Telescope, ELT, of the European Southern Observatory, ESO, with a 39 m primary, planned for Chile in the coming decade, and the Thirty Metre Telescope, TMT, to be built by a consortium including institutions from the United States, Japan, China, India, and Canada, planned for Hawaii in a similar time-frame.

We can round off this piece on telescopes by adding that the optical telescope is still the most direct way to have access to astronomical knowledge and beauty, and as an illustration you have, in Fig. 1.7, an image of the neighbouring galaxy in our local group, the Andromeda Galaxy, (M31) taken by an amateur astronomer, Ivan Bok, who specialises in astrophotography of the highest class, with his own equipment and from within a city. This photograph in visible light over a full wavelength range shows the stars in white through orange, the interstellar dust in dark and patchy lanes, and scattered light in blue in the outskirts. The concentration of stars in the central spheroidal bulge is well presented, as is also the major part of the stellar distribution in a characteristic disc. The disc is warped at its edges, as we will see more clearly when the Andromeda nebula appears again at selected wavelengths in later chapters.

1.1.2 Spectrographs

Much quantitative astronomical information can be obtained from images, and we will see more on this later, but a large part of our understanding of astronomical systems and processes is obtained via spectroscopy. This tells us about the physical and chemical composition of stars, planets, and all distant objects in space. It also allows us to infer physical parameters such as the density and temperature of objects and of the planetary, interstellar, and

a

b

Fig. 1.6 (a) The 200 in. (5 m) Hale Telescope on Mt. Palomar, California, belonging to the Carnegie Institute's Mt. Wilson and Mt. Palomar Observatories. For 27 years, from 1948 to 1975 this was the world's biggest optical telescope. It has an equatorial mount, a fork mount with the telescope between the tines of a huge fork. It needs only one motor, but is at the limit of its size for an equatorially mounted telescope. Credit: Mount Palomar Observatory, California Institute of Technology. (**b**) Interior view of the 10.4 m Gran Telescopio Canarias, (GTC), currently the world's biggest optical telescope.

intergalactic media, which are gaseous. Using spectroscopy, we can measure the velocities towards and away from us of objects, from those in the solar system to the furthest galaxies. We can interpret spectroscopic data using our increasingly complete knowledge of atomic and nuclear physics in the laboratory, which gives us a direct link between the microscopic and the macroscopic, and incidentally shows that matter is essentially the same everywhere in the universe. I will give a very brief look at spectrographs, sufficient to understand how they are used. A spectrograph contains an optical element which can disperse the incoming light according to its wavelength. Traditionally, as first demonstrated with precision by Newton, this element was a glass prism; a schematic diagram of a prism and its dispersion is shown in Fig. 1.8a. But in modern spectrographs the dispersing element is not a prism but a diffraction grating, as schematised in Fig. 1.8b.

A diffraction grating disperses light due to its innumerable parallel slits or grooves; we can see the rainbow effect of fine parallel grooves when light is reflected from the surface of a compact disc or digital video disc (this would have been an ideal illustration 15 years ago, and I trust that many readers will have seen one of these!). The effect was first recorded using bird's feathers by James Gregory in England, shortly after Newton's time, and an artificial grating was made using parallel finely strung hair by David Rittenhouse in Philadelphia in 1785, but gratings were first made systematically for optical purposes in 1821 by Joseph Fraunhofer whom we met earlier as the inventor of the German equatorial telescope mount. Gratings are now present in many technological applications, and their manufacturing techniques have advanced and diversified. They can be made by ruling grooves in a glass surface using a diamond point, by photographic reduction of a set of larger drawn black lines on a transparent background, or by modern holographic methods which are too complex to define or describe here. A result of Fraunhofer's skill in making quality gratings was that the spectrum of sunlight was observable for the first time with spectral resolution high enough to detect the lines produced by different elements in the solar atmosphere.

Fig. 1.6 (continued) At the top, the secondary mirror, of beryllium so that it can be oscillated to make infrared observations, at the bottom the primary, composed of 36 separate hexagonal segments, forming a single mirror when electronically aligned. Its mount is altazimuthal, with axes vertical and horizontal, driven by two computer controlled motors. This system is used by all modern telescopes including small instruments for amateurs. Credits: Gran Telescopio Canarias(GTC)/Instituto de Astrofísica de Canarias (IAC)

Fig. 1.7 The Andromeda Galaxy, M31, our neighbour, in the visible. This is an excellent overview, and includes the Andromeda Nebula's satellite galaxies M32 (close to the disc, left-centre) and M110 (lower right of centre). Credit: Ivan Bok, CC BY 4.0 (https://creativecommons.org/licenses/by/4.0), via Wikimedia Commons

Figure 1.9 shows a spectrum of this kind. The rainbow stripe is crossed by large numbers of dark lines, of different depths and widths. These are known as Fraunhofer lines; in the figure the strongest lines are labelled with the letters A through K, first assigned by Fraunhofer as a practical description. Some of these are still used regularly today by astronomers, who have a conservative streak and often preserve assignations of physical parameters and units first used well over a century ago, to the mild annoyance of other physicists. From the mid-19th century it was realised that the different lines are due to absorption or dispersion by atoms and ions in the solar atmosphere. The best known of these in everyday life are the two closely spaced D lines in the yellow-orange part of the spectrum, a doublet due to neutral sodium, which we see in emission from sodium vapour street lighting (where this still exists). Astronomers still use the terms "H and K" lines for the doublet due to singly ionised calcium, which is found in the spectra of cooler stars.

The diffraction grating is still the basis of most modern astronomical spectroscopy in the visible. Advances have been made to improve spectral

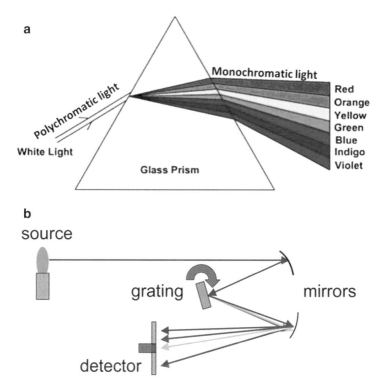

Fig. 1.8 (a) Schematic diagram showing how a prism disperses white light into its constituent wavelengths. Credit: E. Syaodih et al., 2019, J. Phys. Conf. Ser 1280 052051. (b) A schematic spectrograph using a diffraction grating rather than a prism to disperse the light. Credit: Kkmurray, CC BY-SA 3.0 (https://creativecommons.org/licenses/by-sa/3.0), via Wikimedia Commons

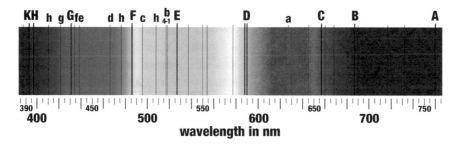

Fig. 1.9 The solar spectrum displayed visually showing some of its strongest lines. Fraunhofer's original simple naming, with capital letters A through K is shown. Astronomers still use the D for the strongest lines due to sodium, and H and K for singly ionised calcium. Credit: Fraunhofer_lines.jpg: nl:Gebruiker:MaureenV, Spectrum-sRGB. svg: Phrood~commonswiki, Fraunhofer_lines_DE.svg: *Fraunhofer_lines.jpg: Saperaud 19:26, 5. Jul. 2005, derivative work: Cepheiden (talk), derivative work: Cepheiden, Public domain, via Wikimedia Commons

resolution (the ability to separate two closely spaced lines or to resolve structure in a single line), and to improve efficiency. It is costly to run a large telescope; observing time is costed in figures of a few thousand euros per night, so ways to improve observing efficiency are of key importance. The lines in a normal spectrum can be thought of as images, in very short wavelength intervals. Another way of seeing this would be to use a very narrow-band optical filter, which can be tuned over a small wavelength range, so that the light emerging would, as scanned show the profile of the line. There are instruments which do this, and which have advantages, particularly for taking spectra of extended objects such as galaxies, as we will see below. But most modern spectrographs use slits and gratings. One of the most efficient ways to observe spectra for an extended object is to use an array of optical fibres in the focal plane of the telescope, which take light from the image and reorganise it to shine into a long single slit. The grating then produces a spectrum for every fibre on the slit, which is a spectrum for every point on the original object. We thus obtain a two-dimensional spectrum using a one dimensional slit, with a great saving in observing time. There are various arrangements using this basic technique, some of which use several bundles of fibres, referred to as IFU's, (integral field units) to image a set of different 2D images at a time, enabling simultaneous spectroscopy of many galaxies in a single astronomical field with a single slit and grating. An example of the optical configuration of a single IFU, or of a complete field of fibres, is shown in Fig. 1.10.

There are a number of arrangements of spectrographs using gratings, which I will mention briefly. One called an echelle uses two gratings in series, but with their rulings essentially perpendicular. One of them, at low resolution, (this is the echelle grating, from the French for a staircase or a ladder because of the shape of its surface) divides the spectrum in pieces according to wavelength, and the other makes a high resolution spectrum of each piece. The result, projected onto a detector, is a spectrum covering a long wavelength range, but compactly organised in a two dimensional array of pieces on the detector. Another type of spectrograph uses a prism with a grating ruled onto one of its faces, known as a grism. This is efficient when observing a field of stars, because it avoids the need for a slit, so the result is a spectrum for each star in a 2D field all at the same time. It is not so useful for extended sources because their spectra can overlap, and the resolution of grisms is not high, to keep the spectra on a single detector. Finally, in this summary of spectrographs we should take a quick look at interferometric spectrographs, which also take spectra over two dimensional fields. Firstly, the Fabry-Perot interferometer whose spectral resolving element is a pair of very closely spaced glass plates. These act as an interference filter, which passes only a very narrow band

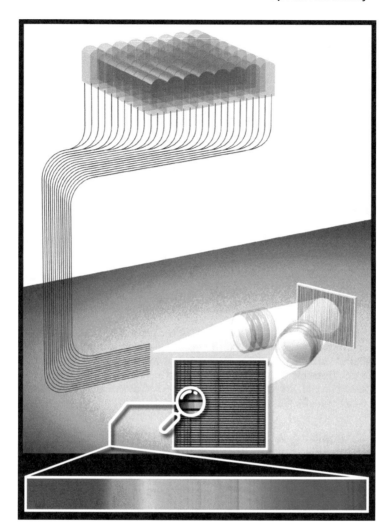

Fig. 1.10 One arrangement of a fibre-fed 2d spectrograph for astronomy. In the focal plane of the telescope a unit of tiny lenses in a square pattern feeds a set of optical fibres, which are reorganized as they feed the light from the square area onto a single slit. The slit forms the entrance to a grating spectrograph, which produces a set of spectra where the inputs from each fibre can be analysed separately, and the spectrum of the square area can be displayed and analysed. The unit is termed an integral field unit, or IFU. Modern spectrographs can have a set of IFU's to produce spectra over quite a large 2d image. Credits: Martin Roth/Inés Bonet-IAC/UC3

of wavelengths. By systematically changing the spacing we change the wavelength which passes through the interferometer. This scanning procedure allows us to measure the variation in intensity across a single spectral line, with very high spectral resolution. These systems are used to make maps of the

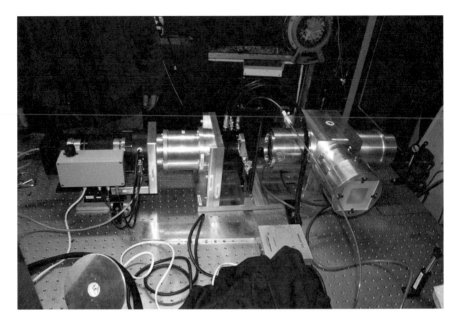

Fig. 1.11 Photograph of the Fabry-Perot interferometer GHαFaS (Galaxy Halpha Fabry-Perot Spectrograph) installed on an optical platform at the Nasmyth focus of the 4.2 m William Herschel Telescope, Roque de los Muchachos Observatory, La Palma, Canary islands. The scale of this small but powerful instrument is given by comparison with the person's hand. Credits: J.Font/Instituto de Astrofísica de Canarias/Isaac Newton Group of Telescopes, La Palma

velocity of complete galaxies, using one of their spectral emission lines, which for the majority of galaxies is the brightest emission line of ionised hydrogen, termed Hα. The maps are the equivalent of the velocity maps of galaxies obtained by radio astronomers in neutral atomic hydrogen with the 21 cm emission line that we will meet in the chapter on radio astronomy.

Figure 1.11 shows a Fabry-Perot interferometer, GHaFaS on the 4.2 m William Herschel Telescope, La Palma and Fig. 1.12 shows the map of velocity across the face of a rotating galaxy as obtained by one of these instruments. Their advantage is that they make fully 2D maps of objects with a very large field on the sky, with very high image resolution and with very high resolution in velocity.

Another type of interferometer which in principle is powerful but has so far been used in a limited way is the Michelson Interferometer. Michelsons have been seen most recently as a core element of the LIGO system for detecting gravitational waves, so the basic configuration can be seen in Fig. 8.5. In LIGO the system is made up of a 45° beamsplitter, which divides the light from the source into two parts of equal intensity, sending one of them to a

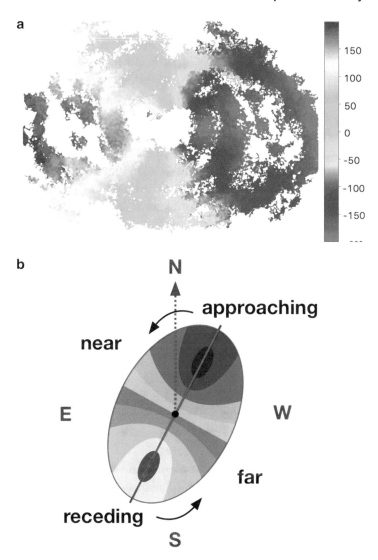

Fig. 1.12 (a) Map of the velocity of rotation of a galaxy taken with the two dimensional spectrograph GHαFaS. The velocity is colour coded, and the pattern shows that the galaxy is spinning on its axis, and tilted with respect to the observer. Velocities are in km/s measured from zero spin at the centre. We can see only the component of velocity directly towards us or away from us. If the galaxy disc were parallel to the plane of the sky we would not detect any of this spin. Credits: Isaac Newton Group of Telescopes, La Palma/J. Font; J.E. Beckman/Instituto de Astrofísica de Canarias. (b) Schematic showing how we interpret the kind of velocity map in (a). A typical disc galaxy spins about a central axis, and is inclined to the plane of the sky. The inlination makes its circular disc appear elliptical. Its rotation velocity starts at zero on the axis then rises till it reaches a flat peak, due to its central bulge, and then flattens out to a constant value in the outskirts. This velocity curve is seen projected along the line of sight to the observer, and a two-dimensional map of the velocity is shown in a stylized way in this figure. Velocities are measured with respect to the centre. Red is a maximum value receding from us, dark blue a maximum value approaching us, and the intermediate velocities have intermediate colours. (a) is a real example of an observed galaxy showing this pattern See also Fig. 2.5b. Credit: IAC-UC3

distant plane mirror by reflection and passing the other part through towards another distant plane mirror. But the type of Michelson used for general spectroscopy has its plane mirrors close to the beamsplitter, and is scanned by systematic motion of one of them. The signal which emerges after the beams have been reflected and recombined is an interferogram. To get the spectrum it is necessary to perform an operation called Fourier transformation of this signal. However, the result is a conventional spectrum, with the advantage of wide spectral coverage and a very big two dimensional field, which enables the observers to map many spectral lines and a wide range of continuum at the same time. Among the few Michelsons in use at leading observatories is SITELLE, on the 3.6 m Canada France Hawaii Telescope. The interferometers require specialised computing skill for their use, but nowadays software which rapidly converts a novice into an experienced user is available for both Fabry-Perots and Michelsons.

1.1.3 Detectors

1.1.3.1 Photographic Plates

For over a century, from the middle of the XIX century to well into the 1960s, quantitative astronomy was based on the use of photographic plates. These have several properties which allowed great advances over the human eye as a detector. The main, and obvious one is their ability to accumulate photons. This property of integration allows observations of fainter, and often more extended, sources than can be seen. In physics and astronomy integration of a signal is generally used to detect faint signals, and although there are limits to detectivity due to various sources of noise, in general terms the longer a signal accumulates, the fainter is the limit to which it can be detected. Photographic plates have another advantage which was not overtaken by modern electronic detectors until the twenty-first century. They are large and can take images which cover relatively large areas of the sky. This made them especially useful for survey work. The famous Palomar Observatory Sky Survey has provided a reference frame for many types of astronomical observations for over half a century after its completion in 1958. It comprised almost 2000 photographic plates, each 14 in. square, covering an area of some 6° by 6° on the sky in two wavelength bands, defined by the sensitivities of the two types of plates used; Kodak 103a-E in the red, and 103a-O in the blue. It was taken using the Palomar 48 in. Schmidt telescope, a reflector with a glass optical corrector plate to optimise the image quality over a wide field. It reached a limiting stellar magnitude of 22, and covered the sky from the north

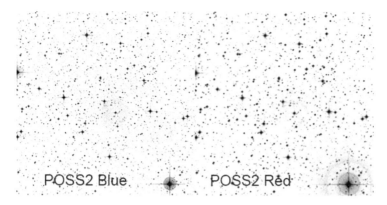

Fig. 1.13 Section of a Palomar Observatory Sky Survey Plate. This survey was in very wide use for over 50 years, for preparing observations (for example finding charts for galaxies and specific stars) and in obtaining basic stellar data. Credit: Mount Palomar Observatory., California Institute of Technology

celestial pole down to below 30° south in declination. It was extended to 42° south in 1966. The plates were digitised between 1986 and 1994, made available on CD-ROM, and can be consulted on a number of websites. In Fig. 1.13 you can see a small section of a pair of red and blue POSS plates to show their general appearance. The importance of size in this kind of work can be understood by comparing the sizes of the earliest CCD detectors, which took images no bigger than three of four arc minutes on a side, only around one thousandth of the size of a POSS plate. Imaging and spectroscopy were both carried out only photographically until the 1960s.

But photographic plates have two drawbacks. The first is their low "quantum efficiency" the proportion of incoming photons they convert to recorded signal. Their technology is essentially the conversión of a silver salt to metallic silver within the emulsion layer, but it requires more than one photon of visible light arriving within a short time interval to produce this conversion. The overall result is that less than 1% of the photons are effectively detected. The efficiency can be enhanced by chemical sensitisation prior to taking the image, but this does not raise it substantially. The second problem is the nonlinearity of the plate. If an astronomer wants a quantitative measurement it is important for the detector to register its detection proportionally to the intensity of the light received. The response of a photographic plate which receives different intensities of illumination is shown in Fig. 1.14.

You can see that for low rates of arrival of the light photons the response rises slowly, there is then a part of the curve which grows linearly with light intensity, but when the detecting silver salt gets close to exhaustion at a given spot, the response flattens out. Measuring this curve on the developed plate

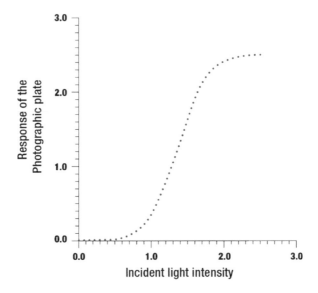

Fig. 1.14 Calibration curve of photographic plate. This is made by shining light of uniform intensity onto the plate which has a filter in front of it, which has sectors from transparent through opaque in steps of increasing absorption. The vertical axis is proportional to the light which has reached the plate, while the horizontal axis shows the response, as measured by the depth of blackness in the image. Once this kind of curve has been measured for a given batch of plates, their response to light from an astronomical image can be linearized. Credit: IAC-UC3

allows the astronomer to linearize images, so that the linearized response is proportional to the incoming intensity at each point. This procedure works, but it introduces a degree of uncertainty, and the range of light intensity for which it can be used is limited. Photographic plates were used both for imaging and for spectroscopy, and did a pretty good job for the astronomical community for many decades.

1.1.3.2   Photoelectric Detectors

The photoelectric effect occurs when a photon hits the surface of a metal, or a photoemissive semiconductor such as gallium arsenide, and succeeds in knocking an electron out of its atoms. If the surface lies within a vacuum tube, the electron can be accelerated away by an electric field and then detected. It is always historically interesting to note that the theoretical explanation of the photoelectric effect won Einstein the Nobel Prize, as his work on relativity was seen as abstruse by conventional physicists. The quantum efficiency of a photodetector of this sort can be as high as 20%, therefore more than an order of magnitude higher than that of photographic plates.

Furthermore, the response to the light intensity is linear, the number of electrons to be detected is proportional to the number of photons. Detection is usually effected by accelerating the emitted electron along a series of positively charged electrodes, "dynodes"; at each collision more electrons are released, so the net effect is that from one incoming photon a signal of up to a million electrons is produced, which is easy to detect in a circuit. This kind of detectors, termed photomultipliers, was used considerably in the 1960s and 1970s for accurate photometric measurements of individual stars, measuring their magnitudes with an accuracy better than 1%. In this period too, the photoelectric detectors were incorporated into imaging devices, such that the detected electrons from a 2D image were both accelerated and focused, using electric fields to accelerate and multiply, and magnetic fields to focus. The result is an amplified image, which can be reconverted to photons by putting a phosphorescent screen at the end of the tube. This image can be hundreds of thousands of times more intense than the initial image at the telescope focus. Photoelectric tubes of this kind have two important advantages over photographic plates: their high quantum efficiencies and the signal amplification; the latter enables recording to be not only fast but linear. This radically improves observations of faint sources, both in image and spectrum. The only drawback is the relatively restricted field size, but for a wide range of observations this is unimportant.

A classical image intensifier tube is schematized in Fig. 1.15; it uses a magnetic field to make a final focused electron image at the end of the tube, which is then linearly converted to an intensified light image by the output phosphor. Much more compact and practical image intensifiers now incorporate a microchannel plate. The electrons from the photocathode are immediately amplified by a 2D grid of very fine tubes with a high voltage across them, organised as a flat plate very close to the photocathode, as shown in Fig. 1.16. At the output of the plate is a phosphorescent screen, and from the screen the amplified light output can be led by a fibre optic bundle to a CCD detector (see the next section) for readout. There is no focusing requirement after the photocathode, because the image configuration is never lost, due to the proximity of all the optical elements. These channel plate intensifiers were invented for use in astronomy, but enjoy much wider use in modern optical and medical engineering.

1.1.3.3 Charge Coupled Devices

The maximum quantum efficiency of a photoelectric detector is 20% so in the last 40 years charge coupled devices, (CCD's) which have much higher quantum

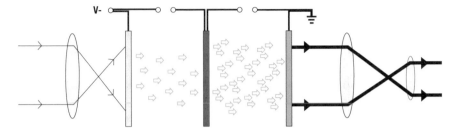

Fig. 1.15 Diagram of an image intensifier tube Photons from a low intensity source, such as an image in a telescope focal plane, enter the objective lens (on the left) and strike the photocathode (grey plate). The photocathode (which is negatively biased) releases electrons which are accelerated to the higher-voltage microchannel plate (red). Each electron causes multiple electrons to be released from the microchannel plate. These electrons are drawn to the higher-voltage phosphor screen (green). Electrons that strike the phosphor screen cause the phosphor to produce photons of light viewable through the eyepiece lenses, or registered on a photographic plate. For precise imaging a designed magnetic field is applied to focus the electrons onto the phosphor. Credit: Eh-Steve, CC BY-SA 3.0 (https://creativecommons.org/licenses/by-sa/3.0), via Wikimedia Commons

Fig. 1.16 CAD design for an Image intensifier with proximity focusing. From top down: entrance window (purple) photocathode (light green) microchannel plate, white, high voltage power circuit (red wires) phosphorescent screen (light green), fibre optic bundle (converging green lines) CCD (square in mixed red green and blue pixels). Credit: Florida State University

efficiencies, have replaced them in almost all optical astronomical instruments. The only places where photomultipliers are still in frontline use are in the huge neutrino detectors which are described in the chapter on neutrino astronomy. The CCD was invented by George Smith and Willard Boyle in 1969 (they shared half the Nobel Prize in Physics for this in 2009, with the other half

Fig. 1.17 A CCD chip on a circuit board with printed read-out leads. Credit: Hamamatsu Corporation

going to Charles Kao who developed the ultrapure glass needed for fibre optics). It works using doped semiconductors which let an incoming photon release an electron from the solid material, which can then be systematically stored and removed within a single circuit. An array of the material gives a 2D pattern of electric charge, which can be transferred to storage while an image builds up. The image can then be read off and used in a computer circuit to reproduce the original pattern of incoming photons, i.e. an image. The quantum efficiency can be well over 90%, and the system is linear over a wide range of photon input rates. A CCD can be used to feed a separate micro-channel plate image intensifier, or can have its own internal electron multiplier circuit. This description of CCD's for astronomy is very basic, there are many variants of detail. For example, many CCD cameras for astronomical use need cooling using either liquid nitrogen or the thermoelectric effect, which reduces thermal detector noise in the image, an important feature for imaging faint objects. Figure 1.17 shows a CCD chip on a circuit board for testing in the laboratory.

As with image intensifiers, the original CCD's were restricted in size. For high image quality the size of an individual pixel should be between 10 and 15 micrometres (μm). This means that for a CCD chip producing an image of capacity 1 Megapixel the chip size is of order 1 cm square. Practical chip sizes of order 3 × 3 cm will then give a 9 Mpixel camera. These figures are normal for moderately priced cellphone cameras. However a chip of a few cm in size will take in only a few arcminutes on the sky at the focus of a normal large reflecting telescope. Although this can be improved somewhat by using lenses to re-image the original image in the telescope focal plane, the basic lower limit to the pixel size means that angular resolution on the sky would quickly be lost as the effective field is increased. The way to reach large fields has been simply to put a number of CCD chips in the focal plane. An example of this is the MegaCam wide-field imaging CCD camera on the 3.6 m Canada-France-Hawaii Telescope at the Mauna Kea Observatory in Hawaii. It comprises 36 CCDs each with a chip containing 2048×4612 pixels, giving a total field of 340 Mpixel to cover a field $1° \times 1°$ square, with pixel size 0.2 arcseconds to give a good sample where the median seeing is 0.7 arcseconds. Even more spectacular is the camera built for the focal plane of the Vera Rubin Telescope (previously called the Large Synoptic Survey Telescope). This 8.4 m telescope, situated in Chile, is designed for comprehensive surveys of the full sky every few nights' observations. To do this it will have a camera with a total size of 3.2 gigapixels, nearly ten times bigger than MegaCam. It is due to start science observations in 2022. The replacement of the Palomar Sky Survey in the CCD age is the Sloan Digital Sky Survey, an immensely fruitful on-going project using a purpose built 2.5 m telescope on an excellent dark sky site at Apache Point, New Mexico. Very deep imaging and also spectroscopy covering the whole sky visible from the site now comprise an archive supplying a wealth of projects and thousands of researchers in many astronomical fields.

I will not devote a section to showing CCD images of astronomical objects, because nowadays virtually all such images are made with CCD's. However I thought it would be interesting to include a selected image, because it shows just what CCD's enable dedicated amateurs to do. It is an image, in Fig. 1.18, of the Veil Nebula, towards the constellation of Cygnus. The observer has put together a mosaic of eight fields to cover this nebula which subtends an angle on the sky bigger than three full moons, and has combined the light emitted by doubly ionised oxygen, in its strong, green spectral line, with that from hydrogen in its strongest visible line, in the red. The nebula is a supernova remnant, which we will come across later in the book.

1.2   Some Essential Results from Optical Astronomy

1.2.1   Photometric Measurements

Photometry is the technique to measure the amount of light received from an astronomical object. A detector sends its signal to a computer, where it is compared with a standard signal to measure the total energy received; dividing by the integration time gives the average power, or luminosity. The standard

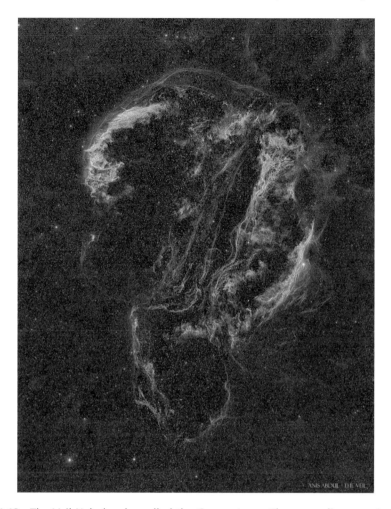

Fig. 1.18 The Veil Nebula, also called the Cygnus Loop. The expanding remains of a supernova which exploded some 10,000 years ago. The diametral size is comparable to three times the lunar diameter. This image shows how a dedicated amateur can obtain beautiful data, especially over wide fields. This required 44 h of total exposure for an eight panel mosaic, and combines light from two spectral lines: OIII in the visual and Hα in the red. Narrow band filters serve to minimise general background light. Taken from a backyard in Austin, Texas! Credits: Anis Abdul/astrobin.com

is a local source, a light, whose power in watts has been measured previously. The comparison requires a knowledge of the beam geometry, the relative angular size of the beam from the comparison source and the beam from the telescope. The power measured that way is the amount of light from the object reaching the detector, per unit time, through the telescope. Given the diameter of the telescope and the amount of light lost in optical transmission through it, we can calculate the power per square metre falling on the Earth from the object. If we know the distance of the object, star, or galaxy, we can calculate more or less directly the power it is emitting, in watts, assuming that it emits isotropically, i.e. uniformly in all directions. The procedure just summarised is the basis of almost all quantitative astronomical measurements made with optical telescopes.

1.2.1.1   Stellar Magnitudes

Nowadays photometry is performed on complete 2D images taken with a CCD camera, such that all the objects in the field are calibrated in one go. But most of the concepts were laid down in the era of photographic plates, and continued through the era of photomultipliers. The measurements are recorded conventionally in stellar magnitudes, which are logarithmic units devised because the response of our eye to a light signal is not linear but close to logarithmic. Basically two stars whose magnitudes differ by 1 magnitude, differ in luminosity by a factor 2.512 which is the fifth root of 100, so that two stars whose magnitudes differ by 5 magnitudes differ in luminosity by a factor 100. The zero point is the observed magnitude of the star Vega at any wavelength (this holds in the visible and near infrared). Directly measured magnitudes are called "apparent" magnitudes, and refer strictly to the light from an object falling on a square metre of the Earth. As explained at the beginning of this section, to convert this to the amount of light emitted by the object we need to know its distance, and a brief explanation of distance measurements will be given later. For the moment, if we know the distance we can change the apparent magnitude into an absolute magnitude, which is a direct measure of the power emitted by the object. The absolute magnitude of an object is the magnitude we would observe if the object were at a standard distance of 10 parsecs, which is 32.6 light years. The Sun has an absolute magnitude of around 5. For historical reasons not easy to defend rationally, objects which are brighter have smaller magnitudes and the brightest objects have negative magnitudes. For reference the total electromagnetic radiation emitted by the sun per unit time is 3.828×10^{26} watts, a measurement now adopted as a definition by the International Astronomical Union. With this definition, a star with absolute magnitude 0 has some 100 times the luminosity of the Sun, i.e. 3.8×10^{28} watts, an object with absolute magnitude -5 would have 10,000 times the luminosity of the Sun, and an object with absolute magnitude $+10$ would have 1/100 of the luminosity of the Sun. When comparing luminosities of objects astronomers often simply express them directly in terms of the Sun's power, so our object with absolute magnitude 0 has 100 solar luminosities, written 100 $L_\odot$.

1.2.2   The Importance of the Cepheid Variables

Historically we can pick out the two most important observations using photometry which underpin most modern observational astronomy. The first was the observation by Henrietta Leavitt of the Harvard University Observatory, that for a specific set of variable stars, the Cepheids, whose luminosity varies regularly with a well defined period, there is a simple relationship between the average luminosity of a Cepheid and its period. She did this using photographic photometry with plates of the two Magellanic clouds, satellite galaxies of the Milky Way, observable from a southern hemisphere observatory belonging to Harvard. She could make this deduction because each "cloud" is much further away from us than its own size, so all its stars can be considered at the same distance, and Henrietta Leavitt understood that she did not need to know the distance to make her deduction. She did not need to know the absolute magnitudes, only the relative magnitudes of her Cepheids. In 1913, a year after she had published her results, the Danish astronomer Ejnar Hertzsprung was able to measure the distances of some Cepheids very nearby to us in the Galaxy, and so to use Leavitt's discovery as the basis for the measurement of the distances of stars. As Cepheids are bright, they can be used to measure distances not only within our Galaxy but to other quite nearby galaxies which has been of fundamental importance for a major part of astronomical research. Members of the Swedish Academy wanted to propose her for the Nobel Prize in 1925 until they discovered that she had died of cancer in 1921 aged only 53. She is an example of an outstanding woman scientist without due recognition in her lifetime.

1.2.2.1   The Hubble-Lemaître Parameter and the Expansion of the Universe

Measuring the distance to an object in space is a fundamental first step for measuring any physical parameter. The Cepheids have played a constant role in distance measurement throughout the century since the work of Leavitt and Hertzsprung. But although they are bright enough to observe in galaxies outside our own, limitations to resolution meant that they could not be reliably observed in the crowded starfields of external galaxies except the nearest. One of the principal goals of the Hubble Space Telescope was to use its resolving power, undisturbed by the atmosphere, to observe Cepheids in galaxies more than 100 million light years (30 Mpc) away, and thus extend a well based distance scale. This objective proved to be of even greater importance because it could resolve the major dilemma about the expansion rate of the universe, the "Hubble-Lemaître constant", H_0. (It would be better to call this the Hubble-Lemaître parameter, because although it takes a constant value over quite a large fraction of the observable universe, it is not constant over very large distances and very large times). From the 1950s until almost the turn of the millennium a controversy existed between two schools of astronomers: whether H_0 took the value of 50 or 100 in units of km/s/Mpc. The measurement needs observations of the "recession" velocity of an object, the velocity with which it is moving away from us, in km/s which is not a difficult measurement, using the Doppler shift of the spectrum, and also the distance to the object, in Megaparsec (Mpc). The latter was producing the headaches. The importance of a good value for H_0 is that the value of $1/H_0$ is an approximation to the age of the universe in any model of a Big Bang expansion. Although many observations pointed at values close to 100, this would put the age of the universe as considerably less than the measured ages of star clusters, and these are based on the quite solid physics of stellar evolution. To place the age of the universe above the ages of the oldest star cluster needed a value close to 50. One of the principal initial aims of the Hubble Space Telescope (HST), shown in Fig. 1.19, was to resolve this problem.

The "Key Project" of the HST , led by Wendy Freedman, was designed to measure the distances of galaxies in the range out to 30 Mpc with unprecedented accuracy by discovering as many Cepheid variables in each galaxy as possible, and determining their periods. In a series of articles published between 1994 and 2000 the team found and measured Cepheids in 18 galaxies, and produced a common table of results for these and another 13. Part of the programme recalibrated the distance to the Large Magellanic Cloud, one of Leavitt's originally observed galaxies, and applied the calibration to the other galaxies. We can see a graph of their results in Fig. 1.20 The final result was a value for H_0 of 71 km/s/Mpc, still with an uncertainty of up to 8 km/s.

This result effectively settled the debate between the advocates of 100 and 50, but did not by itself resolve the crisis of a universe too young for the age of its oldest clusters. However, by combining the Cepheid distances of these galaxies with photometric observations of supernovae of type Ia in the same galaxies, two groups, one led by Saul Perlmutter, and the other led by Brian Schmidt and Adam Riess, measured velocities and distances of much more distant galaxies, out to well over 1000 Mpc (over 3000 million light years) from us. These supernovae have peak luminosities much greater than the Cepheids, and can be observed out to the edge of the galaxy-forming universe. Results from the "High Redshift Supernova Survey" are shown in Fig. 1.21a, and for the "Supernova Cosmology Project" in Fig. 1.21b.

Fig. 1.19 Hubble Space Telescope on orbit, taken from the Space Shuttle just after the fourth and final servicing mission in 2009. Credit: National Aeronautics and Space Administration (NASA)

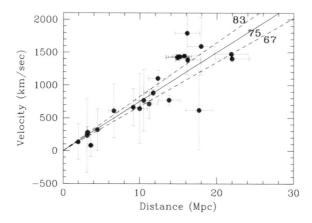

Fig. 1.20 The plot of velocity against distance for the galaxies whose distances were measured in the Hubble Space Telescope Key Project to measure the Hubble-Lemaître constant, Each point represents the distance and velocity of a single galaxy. Distances measured using Cepheid variables in the galaxies. The formal best value for H_0 in this diagram is 75 km/s/Mpc, but refinements using other distance indicators gave a best value of 71 for the work presented by Freedman and collaborators as the final result of the project. Credits: Freedman, W. et al. 2001, Astrophysical Journal, Vol. 553. P.47/NASA

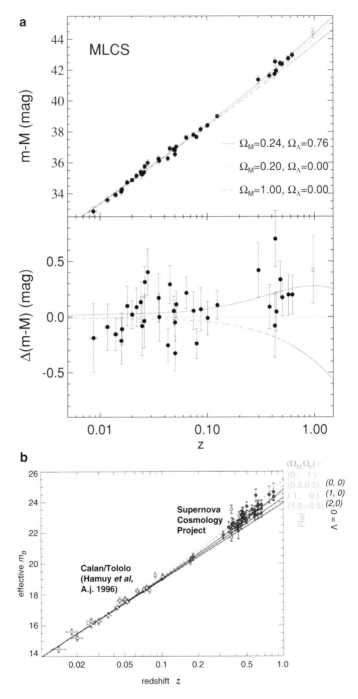

Fig. 1.21 (a) Plot of peak luminosity against distance for supernovae of Type Ia in galaxies as a function of distance, measured in the High-z supernova project The distance is expressed as "redshift" the fractional change in wavelength of spectral lines emitted by the galaxy, due to the expansion of the universe. (a redshift of 0.1 corresponds to a

Both groups found that the universe has not always been expanding at the same rate, and with this information the age of the universe can be older than the simple models with constant expansion. Thus a Hubble-Lemaître constant of around 70 km/s/Mpc does not mean that the universe is younger than its oldest star clusters. Perlmutter, Schmidt, and Riess shared the Nobel Prize for Physics in 2011 for these measurements. Resolving the paradox of a universe younger than some of its stars came at a high conceptual price. The universe is not expanding at a constant rate, but is accelerating. The cause of this acceleration is unknown. It has been termed "dark energy" and its presence is equivalent to some 70% of the mass in the universe. This is at the present time one of the most mysterious components of the universe, and one of the biggest gaps in our knowledge of physics. But we will see in later chapters than this measurement of the Hubble-Lemaître constant is not the last word on the subject.

1.2.2.2   The Hertzsprung–Russell Diagram

The second point of historical importance in photometry is not a single observation but a famous diagram made up of photometric observations. It is the Hertzsprung-Russell diagram (known as the HR diagram), developed independently by Ejnar Hertzsprung and the American astronomer Henry Norris Russell, around 1910. They measured magnitudes of stars through a number of filters at different wavelengths, and plotted their results in a figure which was the original form of Fig. 1.22.

The vertical axis of the figure is the absolute magnitude of the star, found by measuring the apparent magnitude and finding the distance of the star. The magnitude specified in this figure is the visual magnitude M_V, measured through a filter in the most sensitive part of the spectrum for the human eye, in the green, referred to as the V band. The horizontal axis is the difference between the magnitudes measured in the V band and through another filter in the blue, referred to as the B band. The difference is written as B-V in the

---

Fig. 1.21 (continued) distance of 430 Mpc, i.e.1400 million light years). The points are the observations, with vertical bars giving their uncertainties. The lines are the relationships between magnitude and distance for a set of uniform supernovae corresponding to different expansion models for the universe. The best fit is the upper line, which corresponds to an accelerating universe with 76% of the mass in the form of dark energy. The lower diagram shows the deviations from a universe which is neither accelerating or deceleration. From the High z Supernova survey. Credits: Riess, A. Schmidt, B. Et al. 1998 Astronomical Journal, Vol. 116, p. 1009/High-z supernova survey. (b) As figure (a) but for the Supernova Cosmology Project. Credits: Perlmutter, S. et al. 1999, Astrophysical Journal, Vol. 517, p. 565/Supernova Cosmology Project

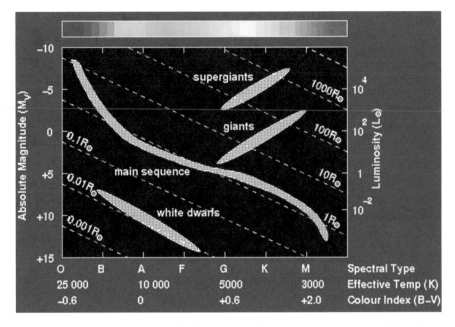

Fig. 1.22 The Hertzsprung-Russell diagram, in which the absolute luminosities of stars are plotted against their temperatures, as found from their colours, which follow the colour bar above the diagram. The absolute magnitude gives the luminosity. The surface temperature, plotted along the axis at the bottom, is measured using the difference in luminosity when the star is observed in two filters, in this case in the blue (called B) and the green (called V, for visual). A third, older way of classifying the temperature is by the letters O through M, also shown along the bottom axis. Credit: Vik Dhillon, University of Sheffield

figure, and is termed the colour index. We can see that it ranges from –0.6 which is the blue end to values approaching +2.0 at the red end of the scale. The vertical axis is the absolute magnitude of the star ranging from –10 at the bright end to +15 at the faint end of the scale. We can also see alternative equivalent scales on the axes. On the horizontal axis we have a scale in temperatures which are equivalent to the B-V colours. On the right hand vertical axis we see the luminosities which are equivalent to the absolute magnitudes on the left hand scale, in units of the luminosity of the Sun.

The principal discovery of Hertzsprung and Russell was that when they measured the magnitudes and the colours of a large number of stars, most of the pairs of values lie on a rather narrow strip going upwards in successive curves from right to left. This strip is called the Main Sequence, and the stars which lie on it, such as the Sun, are called dwarfs. We can also see a number of other areas marked subgiants, giants, supergiants and white dwarfs. This diagram was based purely on photometric observations, but over the years it has become the basis of most of our understanding of stars, their properties and their evolution. When it was first plotted astronomers did not know why most of the stars lie on the Main Sequence, for example. Now we know that a star on the Main Sequence is in equilibrium, so that the power it radiates is balanced by the rate of burning of hydrogen to helium in its core. As hydrogen is the most abundant material in stars, this fuel lasts longer than any other source of energy; the stars stay longer on the Main Sequence than anywhere else in the diagram, so when we look at a population of stars, we find more of them on the Main Sequence than anywhere else. Stars higher along the Main Sequence have more mass than stars lower down. All the features of the HR diagram, the properties of the dwarfs, giants, supergiants, and so on, are now explained in terms of the different nuclear processes which fuel the stars, and which lead them to have a given magnitude and a given colour at a specific time during their individual evolution. Let us consider a simple example of a physical inference from the diagram. We can take three stars, one of absolute magnitude 6, one of absolute magnitude 0 and one of absolute magnitude –6, with all three having the same colour, for example a value for B-V of +0.6. The fact that they have the same colour means that they all three have the same surface temperature, of just under 6000 K. But the fact that they have different magnitudes means that their power output is different, they have different luminosities. The star of magnitude +6 is a little fainter than the Sun, while the star with magnitude 0 is almost a hundred times brighter, and the star with magnitude –6 is 25,000 times brighter. If we ask how this can happen, it must mean that the two brighter stars must be bigger than the faintest one. The second must have a diameter 10 times that of the Sun, and the third one a diameter 150 times that of the Sun, hence the names giant and supergiant. Once we know this, those who explain the properties of stars must have theories which can account for these differences. Most of stellar astronomy is based on the physics of stellar structure and evolution, which can be condensed into an understanding of the HR diagram.

1.2.2.3 Two-Dimensional Photometry of Galaxies

To show what is being learned using two dimensional photometry of galaxies we will look at studies of the faint outskirts of disc galaxies, because these give interesting clues about how galaxies form, and because they push technique to the limit. Deep, long exposure images of galaxies were made towards the end of the photographic plate era, back towards the end of the 1970s. The images of edge-on galaxies looked particularly interesting because they showed a sharp cut-off, or truncation at the limits of the disc.

a

Fig. 1.23 (a) Edge-on galaxy NGC 4565 taken by combining exposures through three colour filters taken with a CCD camera Credit: Ken Crawford, Rancho del Sol Observatory. (b) Isophote plot of NGC 4565 from a publication in the 1981 made with a photographic image The isophotes represent the surface brightness, highest at the centre of the galaxy and fading down towards the edges. Credits: P. van der Kruit, L. Searle, 1981, Astronomy & Astrophysics Vol. 95 p. 105

In Fig. 1.23 we can see a modern CCD image of the edge-on galaxy NGC 4565 and also a map of the isophotes, lines of equal brightness, published from a photographic image in 1981. The theorists in the 1980s began to work on models showing how truncations could be produced. Three physical effects were suggested. The first was, quite simply, that there is a threshold for the gas density in the interstellar medium needed to form stars, and outside a certain radius of a galaxy disc the gas density drops below this threshold, so star formation is truncated. The second was that gas distribution in a forming galaxy is affected by its rotation, which puts a sharp limit on the extension of the gas, and hence of the star formation in the galaxy which forms. The third

b

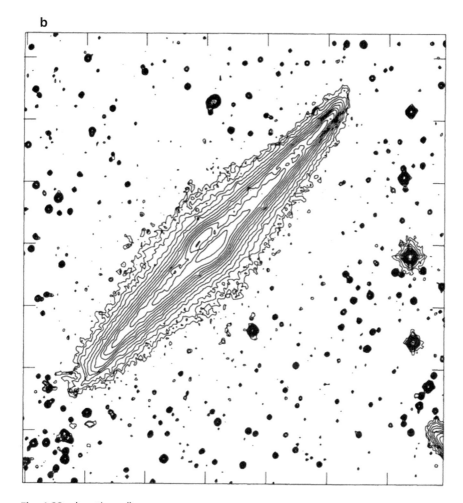

Fig. 1.23 (continued)

was that magnetic fields control the gas distribution, and stars stop forming at a radius where the magnetic field of the intergalactic medium is bigger than that of the galaxy. When CCD's were in general use observers began to take deep exposures of face-on galaxies which was then possible as CCD images can go to fainter isophotes. They expected to make improved observations of truncations, because their discs are not foreshortened as they are in edge-on galaxies. To their surprise they did not find truncated discs. They found instead three different types of behaviour in the light profiles of the outer discs, Type-I where the surface brightness gradually and evenly fades until it is too faint to detect, Type-II where at a given radius the surface brightness does fall off more quickly, but gradually and not sharply as would be the case for a truncated

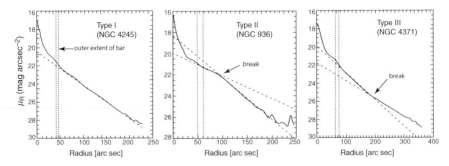

Fig. 1.24 Three representative types of surface brightness profiles of disc galaxies. The brightness in magnitudes per square arc second at a given radius from the centre of the galaxy is plotted against that radius. The brightness at a given radius is an average taken around an annulus of the disc at that radius. Notice that there is no sharp cut-off in the brightness, i.e. no sudden dearth of stars at the edge, rather a steady decline, even for the Type II galaxies. Credits: Erwin, P.E, Beckman, J.E., Pohlen, M., 2008, Astronomical Journal, Vol. 135, p. 20

disc, and Type-III, where at a given radius the surface brightness does keep fading down, but less quickly that in the inner disc, a sort of "anti-truncated" behaviour. Illustrations of the three types of profiles are shown in Fig. 1.24.

Clearly the efforts by the theorists to explain truncations via models of the evolution of galaxies based on the photographic evidence were not likely to succeed. To understand the process of galaxy building at the edge using deep imaging needed new theoretical approaches as well as more observations. It is completely implausible that edge-on galaxies have really sharply truncated discs, while face-on galaxies do not. Some sort of projection effect must be in play. This was explored quite recently by Alejandro Borlaff and collaborators. Starting with the observation that the disc of the Milky Way fans out into a flared shape at the edge, they made numerical models showing how the light profiles of galaxy discs would appear to us if they were flared, when observed edge-on and face-on.

One of their results is shown in Fig. 1.25, where we can see that the same flared galaxy would show a steep fall in surface brightness if observed edge-on, but much less steep if observed face-on; this does not hold for a galaxy whose disc is not flared, as we can also see in the figure. We can also see that the isophote map of the edge-on flared galaxy looks more like the map of the disc of NGC 4565 in Fig. 1.20 than does the map of the unflared galaxy. We should add that this flare effect on the observations does not mean that some of the physical effects mentioned above are inoperative. The star formation rates and densities at the edges of discs certainly fall with increasing radius, and eventually the disc will either peter out or stop more abruptly, but this requires even deeper imaging to detect well.

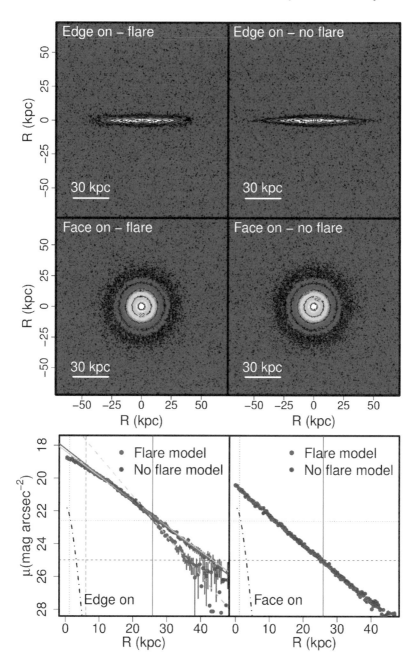

Fig. 1.25 The results of modelling the disc of a galaxy whose stars flare outwards at the disc edge. A flared disc observed edge-on will show a Type-II profile as measured along a line running through the centre of the image, but will show a Type-I profile if observed face-on. A disc without a flare will show a Type-I image observed from either direction. Credits: A.S. Borlaff et al. 2016, Astronomy & Astrophysics Letters, Vol. 591, p. L7

The Type III profiles are more puzzling. Why should the decline in brightness with radius become less outside a given radius? The theorists think that this is because a Type-III galaxy has absorbed another galaxy, and in the merger the redistribution of the stars gives rise to the brightness distribution seen as a Type-III, and there are models to show that this is plausible. As evidence grows that the evolution of galaxies on cosmological timescales has involved mergers between galaxies, observations over increasing timescales are needed to explore the properties of the Type-III galaxies. This is done by observing at higher redshift, and comparing with local galaxies. In order to obtain images good enough to measure the brightness profiles at significant redshifts the best source of observations up to now is the archive of the Hubble Space Telescope (HST), and the deepest images of galaxies ever obtained by the HST are in the Hubble Ultra-Deep Field, obtained by exposing a single small area on the sky for a total of almost two million seconds (over 22 days). This field has been used for many projects, mostly to identify different types and properties of galaxies by spectroscopy from the largest ground-based telescopes, and among other objectives to measure their redshifts.

Figure 1.26 shows the results of a specific treatment of this field by Borlaff and collaborators, designed to make correctly linear images down to the faintest magnitudes, almost to 33 magnitudes per arcsecond squared. So far the properties of the Type III galaxies found at redshift 0.6, some 5000 million years ago, are almost identical to the profiles of local galaxies, but almost four times brighter. Work is in progress to find Type III galaxies at redshift 1, i.e. some 7000 million years ago, to explore their properties, but progress is slow because the observers are pushing the angular resolution even of the Hubble Space Telescope to its limit when observing the profiles of these galaxies, and only the exquisite treatment of the photometry implied in Fig. 1.23 will allow the light profiles to be measured reliably. Further progress will need new generations of observations in the near infrared from projects such as the James Webb Space Telescope.

Our last look at imaging in the optical wavelength range is at galaxies where the precise geometry has been lost, but which can give us a vast amount of new information combining imaging with spectroscopy. These are very distant galaxies which have been gravitationally lensed so that we can see them much more intensely than if their light had simply been emitted to us and spread uniformly through space. Gravitational lensing is one of the properties of space-time predicted by Einstein's theory of General Relativity, which predicts that light should take curved paths as it travels close to massive bodies. It was first detected in 1919 in the famous experiment to measure the gravitational shifts in the apparent positions of star images close to the Sun during an

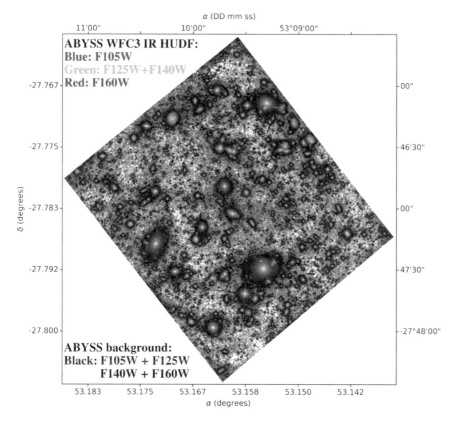

Fig. 1.26 A portion of the Hubble-Lemaître Ultra Deep field specially reduced to produce a true linear image down to the faintest magnitudes in order to make surface brightness profiles of galaxies observed at redshift 1 (approximately 7000 million years ago, at around half the age of the universe) with the specific aim of detecting and analysing Type-III profiles. This image is the deepest image of the Universe ever taken from space. The coloured galaxies are surrounded by dark regions which are representations of the light from the outermost parts of their discs. These regions are presented as dark to bring out their shapes against the sky background which is white. Credits A.S. Borlaff et al. 2019 Astronomy & Astrophysics, Vol. 621, p. 133

eclipse. In recent decades it has become an increasingly used tool to study distant phenomena.

To see how gravitationally lensed galaxies appear when observed take a look at Fig. 1.27, which is an HST image of the "nearby" galaxy cluster Abell 370, in which arcs of light emitted by much more distant galaxies behind the cluster, are seen. These arcs have been produced by the gravitational lensing effect of the cluster itself. Numerical estimates of the mass needed to produce these arcs are among the evidence of the existence of a dominating quantity of

Fig. 1.27 Hubble Space Telescope image, combining infrared and optical data, of a set of gravitationally lensed distant galaxies, behind a relatively nearby cluster Abell 370. This cluster is, in cosmological terms, quite close to us, at a redshift z of 0.25, corresponding to a distance of 3000 million light years. The circular arcs are the focused light from galaxies at much greater distances, ranging from z = 3 to z = 7. They are used to infer the structure of the dark matter within the cluster as shown in Fig. 5.16. Crédits: NASA/ESA/Jennifer Lotz y the HFF Team(STScI)

dark matter in clusters of galaxies, even though the mass of ordinary baryonic matter is itself dominated by the hot gas not seen here, but detected in X-ray observations. In the chapter on X-ray astronomy we see the image of this cluster in X-rays and the inferred distribution of the dark matter derived by analysis of the gravitationally lensed arcs. The brightest of these arcs can be observed spectroscopically by 10 m class telescopes on the ground to give valuable information about their stellar and gas content. This is how we are learning about the properties of galaxies at redshifts of 5 and greater, going out to redshifts of over 10, when the universe was less than 500 million years old, (i.e. well over 13,000 million years ago).

1.2.3   Optical Spectroscopic Observations

Most of the work in astrophysical research is still based on optical spectroscopy. So much so that even to summarise 1% of it would be well beyond the scope of this book. So I have simply chosen a few observations, using purely personal criteria, which I think are of particular interest.

1.2.3.1   Abundances of the Chemical Elements

One of the basic overarching aims of astronomical spectroscopy has been to measure the relative proportions, normally referred to as abundances, of the chemical elements in different sites and at different epochs in the universe. With the development of the nuclear physics of energy generation in stars in the 1950s, synthesised notably in a famous paper by Burbidge, Burbidge, Fowler, and Hoyle in 1957 (affectionately called B^{2}FH in the astronomical community), the theme of nucleosynthesis of the elements took centre stage. The different processes by which elements fuse within stars, releasing energy and producing heavier elements, occur in different proportions in stars with different masses and at different stages of their life histories. Coupled with the theory of how stars with different masses are structured and develop, we have the basis for exploring not only the histories of stars but also of galaxies via their "chemical evolution". We can go further and compare the abundances of the elements observed in stars and the interstellar medium with those observed on Earth, and in the meteors, comets, asteroids and planets of the Solar System. The common tool is spectroscopy, mainly optical, but as techniques have developed, extended progressively into the infrared and ultraviolet.

The basis of the use of abundance observations to trace evolution is the periodic table of the elements, tagged with the different processes and the different kinds of stars which produce them, shown in Fig. 1.28 In the Big Bang cosmologies, within the first few minutes after the Big Bang the fundamental and lightest element, hydrogen, was the first to form from the "primordial soup" and while the universe was still hot enough for nuclear reactions to occur everywhere, just over 20% of the hydrogen fused into helium, which

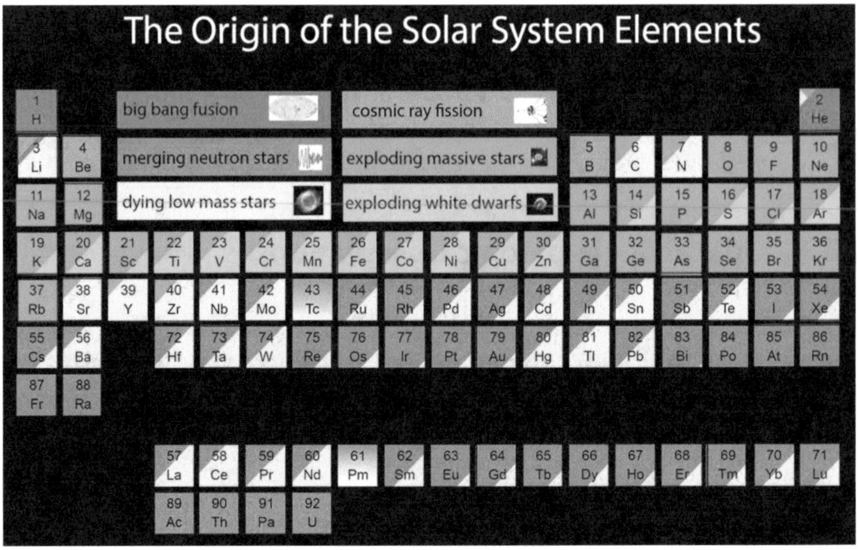

Fig. 1.28 The origin of the elements. Each element is tagged with a colour code showing its site of origin. The great majority of the elements were built up within stars at different stages in their evolution. Only hydrogen, most helium and a tiny amount of lithium were produced in the big bang, while some lithium, and all of beryllium and boron were produced by the action of cosmic rays on heavier elements in the interstellar medium. Credits: Jennifer A. Johnson/NASA/ESA

has been the principal source of helium until the present day. A very tiny fraction of the Lithium in the universe also formed in the Big Bang.

1.2.3.2   Abundance Measurements in HII Regions

Spectroscopy is used to measure the relative proportions of the elements, their "abundances", and abundance is defined as the proportion of the given element in a star, or in the interstellar medium relative to Hydrogen, which is always used as the reference element. The technical details of how to use spectra to measure abundances will not be tackled in this book, but you can find references in the "Further Reading" at the end of the chapter. However, an important point about these measurements is that they may be made in the spectra of individual stars, within the Milky Way or galaxies near enough to separate individual stars, they may be made in the spectra of whole galaxies or parts of galaxies, or they may be made in the interstellar medium. One of the most usual ways to make element abundance measurements is to use the emission lines from a given element in the hot ionised gas around a cluster of hot stars, or an individual hot star. These zones or ionised gas are called HII

Fig. 1.29 The Orion Nebula, our nearest HII region. The hot stars are blue-white ; the emission from ionised hydrogen is in red, the dust produced by stars which have already exploded as supernovae is in the dark lanes and patches. Credit: Stellina, app-controlled telescope from French start-up Vaonis, in the Business Incubation Centre-Sud, of the European Space Agency (ESA)

regions, from the spectroscopic notation that for a given element, call it E, the neutral form is termed EI, the singly ionised form is termed EII, and the doubly ionised form (which has lost two of the original electrons) is termed EIII, etc. Around the hot stars, classified as types O or B, there are regions where the radiation from the stars ionises much of the hydrogen surrounding them, so they are called HII regions. The emission lines from HII regions are relatively easy to detect and measure, both in our Galaxy and in other galaxies, extending out to large redshifts. So they are widely used to measure the proportions, the abundances, of the different chemical elements in the universe in general. The closest and best studied HII region is the Orion Nebula, which has been subject to very detailed observations both from the ground and from space (Fig. 1.29).

1.2.3.3 Rotation Curves of Galaxies: Dark Matter

Galaxies emit visible light from their stars and also from the ionised gas in their interstellar media. Stars have a multitude of spectral lines, in absorption, and the composite spectrum from the stellar population of a galaxy shows these lines plus the emission lines from the ionised gas in the interstellar medium, in most disc galaxies with star formation. Both types of lines allow us to measure the velocity along the line of sight to all the points on the galaxy. From this information we can produce a plot of the rotation velocity of the galaxy as a whole, and how it varies with radius from the centre of the galaxy. This is called a rotation curve.

In Fig. 1.30a you can see two-dimensional maps of the velocity distribution across the face of a galaxy, and in Fig. 1.30b the rotation curve which has been derived from it. The range of rotation velocities, up to around 220 km/s, is typical of galaxies in the mass range of the Milky Way. The rotation curve of a galaxy is somewhat similar to a graph of the velocities of the planets in their orbits around the Sun, but with a difference. In the Solar System almost all the mass is concentrated in the Sun itself, and under these circumstances the velocity of a planet around its orbit is lower the further away it is from the centre, i.e. from the Sun. Quantified, this is the famous Kepler's famous third law of planetary motion, which Newton used to derive his law of universal gravitation. But as the distribution of matter in a galaxy is not so concentrated towards the centre, one would expect a much slower decline in circular velocity with radius that for the solar system. Looking at Fig. 1.30b we can see that there is no obvious decline with radius at all for NGC 3433. It was by combining optical rotation curves with those measured using atomic hydrogen emission at 21 cm radio wavelength that Vera Rubin in the 1970s obtained rotation curves of galaxies out to large radii in their discs, radii at which the combined density of the stars and the gas is very low. To her surprise, and that of most astronomers, the velocities of most galaxies did not fall with increasing radius, even well beyond the point where Newton's law of gravity indicates that they should. She concluded either that the law needed modification, or that there is a great deal more matter in the outer parts of galaxies than we can easily detect. Personally she preferred the first option, which has been explored by theorists under the heading of MOND (modified Newtonian Dynamics), but the consensus among most astrophysicists is that the second option, that of a major quantity of invisible "dark matter" is much more probable as it is supported by a variety of further observations. The concept of dark matter is now deeply rooted in astrophysical research, and is incorporated into the cosmological models which aim to explain the origins of galaxies and their clusters. The current view is that there is some five times more mass in the form of dark matter than in "normal" baryonic matter (made up of protons and neutrons plus electrons). There are candidates in the "zoo" of particle physics for the components of this dark matter, but so far they have not been detected. We will meet with this dark matter mystery in other chapters of this book.

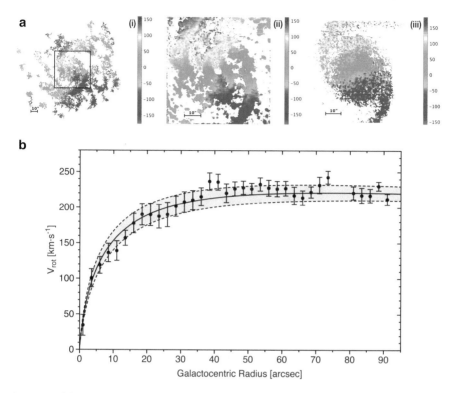

Fig. 1.30 **(a)** Two-dimensional velocity maps of the spiral galaxy NGC 3433 (i) Map of the major part of the disc from its interstellar emission, made with the Fabry-Perot spectrograph GHαFaS on the 4.2 m William Herschel Telescope, La Palma, (ii) Map of the part of the disc within the black square in (i) from gas emission, with the MUSE 2D fibre-fed spectrograph on the VLT in Chile, (iii) map of the velocity measured from the stellar spectra with MUSE. The velocities are colour coded so that for this, a typical rotating galaxy disc, the blue zones are moving towards us, the red zones are moving away from us, and the colour trend across the disc represents the global effect of rotation as projected along the line of sight. **(b)** The rotation curve of NGC 3433 made by averaging over the 2D map in (a). The points are the average circular velocity at a given radius, the black curve is a smooth polynomial fit to the data, and the green shaded area marks the statistical error limits of the curve. Credits: Beckman, J.E. et al. 2018, Astrophysical Journal, Vol. 854, p. 182/IAC/ING/European Southern Observatory (ESO)

Further Reading

Book "Eyes on the Sky" by Francis Graham Smith. Oxford University Press, 2016. Discusses astronomy from the point of view of its telescopes, starting in the optical and covering the electromagnetic spectrum.

Book "Light, the visible spectrum and beyond" by Kimberly Arcand and Megan Watzke, Black Dog & Leventhal, 2015 An introduction to the use of radiation at different wavelengths in astronomy.

Multiwavelength astronomy, on-line from Chicago university: http://ecuip.lib.uchi-cago.edu/multiwavelength-astronomy/index.html. This covers optical, infrared, ultraviolet, X-ray and gamma-ray astronomy in simple terms but by some leading experts. It has a historical base. Particularly good for early high-school students

Optical Astronomical Spectroscopy. Author C.R. Kitchin Publisher: The Institute of Physics. An introductory textbook for those without specialist knowledge.

2

Radioastronomy

2.1 The Beginnings of Radioastronomy

The first of the "new astronomies" which heralded the modern age in which the range of observations in the electromagnetic spectrum was widened by over fifteen orders of magnitude, was radioastronomy. The beginnings were made in the 1930s by Karl Jansky and Grote Reber in the United States. Jansky, working at Bell Laboratories, was trying to find the source of the noise being detected on radio receivers operating at frequencies around 20 MHz (a wavelength close to 15 m). He built a steerable antenna and took measurements in different directions over the sky. Figure 2.1 shows Jansky with this, the first radiotelescope.

He found that the source was from outside the atmosphere, but did not detect emission from the Sun. It had a generally disc-like distribution, and its peak intensity coincided with the position of the centre of the Milky Way. Astronomers in general did not show much interest, but the radio engineer and amateur astronomer Reber built a 9 m parabolic antenna in his garden and in 1939 made a map of the Galactic plane at 160 MHz (1.8 m wavelength) which also showed a peak near the Galactic centre, and also found a source towards the constellation of Cygnus, now known as Cygnus A, which is an external galaxy, and one of the strongest radio sources in the sky. These were the pioneers (Fig. 2.2).

In the early 1940s two developments took place which were to lay the foundations of radioastronomy as we know it today. Firstly radio receivers at short wavelengths were developed in laboratories dedicated to radar as a key to the war effort, and with these the Sun was detected by scientists in the

© Springer Nature Switzerland AG 2021
J. E. Beckman, *Multimessenger Astronomy*, Astronomers' Universe,
https://doi.org/10.1007/978-3-030-68372-6_2

Fig. 2.1 Karl Jansky with his radiotelescope, the first ever made. It could be rotated on a set of wheels from a Ford Model T car to observe different directions on the sky. Although this does not look much like a telescope at long wavelengths one may not need a filled aperture antenna, and many open work radiotelescopes have been built since then. Credits: National Radio Astronomy Observatory (NRAO). The NRAO is funded by the National Science Foundation (NSF)

United Kingdom and the United States over a range of wavelengths up to 3 cm, both from the disc as a whole, and from individual sunspot areas. This heralded the introduction of the technical elements of radar into astronomy, giving ever more sensitive and accurate devices (Fig. 2.3).

On the side of theory Henk van de Hulst at Leiden in the Netherlands predicted in 1944 that atomic hydrogen in interstellar space should be detectable via an emission line at 21 cm wavelength (1420 MHz in frequency). As hydrogen was expected to be very abundant in the universe this prediction was of great importance; the line was first detected in 1951 by Harold Ewen and Edward Purcell at Harvard University, and confirmed by Dutch astronomers Muller and Oort, (collaborating with van de Hulst), also in 1951. The importance of detecting a spectral line is that it gives two types of information: the strength of the line gives a measure of the quantity of material producing it, and the wavelength of the line measures the velocity of the emitting material with respect to us, the observers. So when the 21 cm line of atomic hydrogen was detected astronomers began working to make a map of the hydrogen in the

Fig. 2.2 Grote Reber's radiotelescope, precursor of many modern astronomical antennae. Credit: NRAO The NRAO is funded by the National Science Foundation (NSF)

Galaxy. This way of mapping gave advantages over previous observations in visible light. A key advantage was that hydrogen is ubiquitous in the zone of the Galactic plane, so it can give a more continuous map than the stars. Another is that the dust which is present in all interstellar gas is essentially transparent at 21 cm wavelength, whereas it is strongly absorbing in the visible, and blocks our line of vision close to the plane. So in principle a 21 cm map could cover the whole Galaxy, whereas along lines of sight close to the plane our optical

Fig. 2.3 Henk van de Hulst, Dutch astrophysicist who predicted that the 21 cm line of atomic hydrogen from interstellar space could be observed, and would help to map the Galaxy. Credits: ANEFO/Rob Bogaerts

observations are limited to distances of less than 10% of the Galaxy diameter. Astronomers in the Netherlands and Australia combined to make the best possible map of the interstellar hydrogen across the whole galaxy, producing the famous diagram shown in Fig. 2.4. One of the important features of making a 21 cm map like this is that as well as the intensity of the emission line, you measure its velocity, so that you obtain a two-dimensional velocity map of the part of the Galaxy you have observed. From this the observers produced the rotation curve of a part of the Galaxy, shown in Fig. 2.4b. But note that observing within the Galaxy puts a different type of limitations on the result. In order to make the map it is necessary to derive the distances of all the hydrogen clouds, and this is done using a model of the rotation of the galaxy, in which the velocity of a cloud with respect to the observer can be used to infer its distance. You can see that there is a blank conical zone where this velocity cannot be well measured because the gas in the zone has no velocity component towards or away from us, and that is the only measurable velocity. This specific difficulty in mapping our Galaxy in HI is normally much reduced in external galaxies.

It is interesting to see that the cone of absence in the Milky Way map is not present in the equivalent map of the Andromeda Galaxy, M31, our nearest neighbouring large galaxy. Both the hydrogen column density and the velocity can be obtained in two dimensions, using the assumption that rotation within the plane of the galaxy disc dominates other motions. We can see the results of this in Fig. 2.5, and the colour coded velocity distribution in the

→

Fig. 2.4 (continued) bined observations from radiotelescopes in the northern and southern hemisphere **(b)** Curve of the velocity of rotation of the disc of the Galaxy v. radial distance from the centre, (usually called the "rotation curve") obtained from the velocity map derived using the 21 cm data depicted in R 4a. The points are from the observations, and the curve is an envelope fit to the data. Credits:J. Oort et al.Monthly Notices of the Royal Astronomical Society (MNRAS), 1958 Vol.118, p. 379. Published courtesy of the Royal Astronomical Society

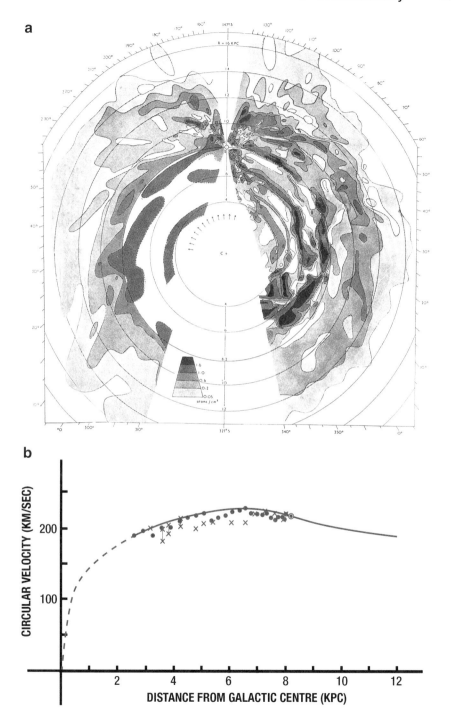

Fig. 2.4 (a) First comprehensive map of a major part of the disc of our Galaxy observed using the 21 cm line of atomic hydrogen by Oort, Kerr, and Westerhout, published in Monthly Notices of the Royal Astronomical Society Vol 118, p. 379, 1958. They com-

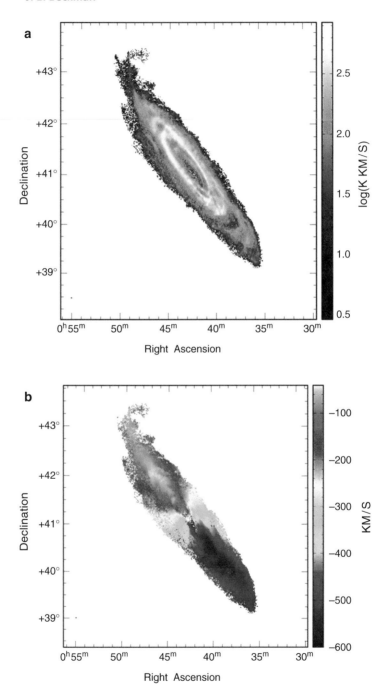

Fig. 2.5 (a) Map of the atomic hydrogen in our Local Group Galaxy M31, obtained using the 21 cm line with the Dominion Astrophysical Observatory radio interferometer plus the 26 m single dish. (b) Map of the velocity of the atomic hydrogen in M31 from the data set used for the map in a. The velocity observed is the component directed towards us, the observers, and is typical of what we observe for a rotating disc

disc is equivalent to the colour coded distribution obtained using emission from ionised hydrogen at optical wavelengths shown in Fig. 1.12, although the disc of M31 is much more inclined to our line of sight.

2.2 Radiotelescopes: Single Dishes and Interferometers

Once astronomers had begun to detect astronomical sources in the radio wavelength range they began to build special aerials, now generally referred to as antennae, or more colloquially as "dishes" to make the observations, as well as working on increasingly sensitive receivers. These dishes fall into the general category of radiotelescopes, although as we will see there are a number of different designs for different purposes. The United Kingdom and the Netherlands were among the key pioneering countries in developing these radiotelescopes, as well as the United States and Australia; the Russians also made important contributions. In the United Kingdom in the 1950s two major telescopes were built. One was a conventional single dish, the famous Jodrell Bank radiotelescope built by Bernard Lovell between 1951 and 1957, and shown in Fig. 2.6.

This fully steerable 250 ft (76.2 m) dish was a triumph of enterprise. Lovell, and the University of Manchester ran into serious debt during the construction. They were saved by the launching of the first satellite, Sputnik 1, by the Russians in October 1957. The Jodrell Bank telescope (later termed the Mark 1, and finally named after Sir Bernard Lovell) was able to track the third stage of the carrier rocket using radar. The ability to track satellites, especially those moving further away from Earth than mere close orbit was a unique facility, and significant funding was received in payment for tracking satellites, both from the US and the USSR, which helped to resolve the debt crisis. Jodrell Bank has been an international centre for radioastronomical research ever since that period. The large collecting area of single dish telescopes makes them particularly useful for observations of fainter objects. A number of large fully steerable single dishes, in the style of Jodrell Bank, have been built over the years. These include the 100 m dish at Effelsberg, belonging to the Max Planck Institut für Radioastronomie in Bonn, Germany, and the 100 m Green Bank telescope in West Virginia, belonging to the National Radioastronomy Observatory of the United States, which is the world's largest fully steerable radiotelescope.

Fig. 2.5 (continued)   galaxy when the internal rotation in the plane of its disc is the dominant velocity. A schematic form of this diagram is shown in Fig. 1.12b. Credits: C. Carignan et al./Astrophysical Journal 2009, Vol. 705, p.1395

A fully steerable telescope can, of course, observe any point on the sky in its hemisphere. But the engineering difficulty of building a fully steerable telescope much bigger than 100 m in diameter which does not deform as it is steered around the sky has meant that bigger radiotelescopes have been built as fixed. For many years the biggest single dish radio telescope in the world was at Arecibo in Puerto Rico. This was constructed in a bowl shaped valley which not only supports the 305 m dish, but also gives considerable protection against stray radio noise from terrestrial sources which could vitiate the observations. The telescope can be pointed effectively to a limited range of variable positions on the sky by moving its receiver, which is suspended over the dish from three major pylons, as own in Fig. 2.6.

Epitaph for a telescope.
It is sad to have to report that while I was finishing this book the great radio telescope at Arecibo was fatally injured. After suffering Caribbean storms over the summer months, on August 10th (2020) one of the main support cables snapped, and on November 10th a second cable snapped. Pieces of the central platform carrying the foci fell onto the main dish producing considerable damage to an already weakened frame. The experts called in by the National Science Foundation made an assessment that it

Fig. 2.6 The Lovell telescope, originally the Mark 1 radiotelescope at Manchester University's Jodrell Bank observing station in Cheshire, U.K. one of the earliest large single dish radiotelescopes. Credit: The University of Manchester

would be too dangerous to try to recover the instrument, and the NSF announced its demise. It had been badly affected by a magnitude 6.4 earthquake in 2014, but could then be restored to working order. Among the programmes badly affected are the ongoing radar studies of asteroids, particularly of near Earth asteroids, which no other telescope (including the Chinese FAST radio telescope, see below), is equipped to do. It will be harder to find a way to substitute the Arecibo telescope than it was to make a replacement for the 91 m Green Bank radio telescope, West Virginia, destroyed in a storm in 1988.

The Arecibo telescope was the largest in the world for decades, but has since been overtaken by two extremely large instruments. The RATAN telescope, at Zelenchukskaya in the Caucasus, has a maximum diameter of 600 m and a rather complex system. It is not a continuous filled bowl, but a set of mirrors in concentric rings, which contribute by sending their signals to a central conical receiver. The biggest single dish radiotelescope, completed in 2016, is FAST, the 500 m aperture spherical radio telescope, at Kendu, Guizhou province, China, shown in Fig. 2.7. This is a more modern version of the Arecibo dish, and can be effectively pointed over a moderate area of sky near its zenith by moving the detector at its focus, and by moving the separate panels of the

Fig. 2.7 The Arecibo radiotelescope in Puerto Rico, for nearly 50 years the world's largest single dish radiotelescope. The dish, of diameter 305 m, is firmly fixed in a karst valley formation. The telescope can scan the sky in a cone of full angle 40° by adjustment and movement of the secondary mirror and detector seen suspended by steel cables above the middle of the dish. . As I write this news has come in that this telescope has suffered irreparable damage and is being closed down Credit: NRAO-NSF

main dish, controlled by computer. One of its main tasks is to work on pulsars in globular clusters of stars (Fig. 2.8).

One of the limiting factors for radiotelescopes is their angular resolution on the sky. This is the ability to resolve objects, which may be to separate two neighbouring point sources, such as stars, or to measure structure in a single extended object, such as a galaxy or an emission nebula. The resolution of any telescope depends on two parameters, the wavelength of the light it uses, and its diameter. To be precise it depends on the ratio of the wavelength to the diameter, so that the resolution is improved either by using a shorter wavelength or by using a telescope with a larger diameter. Of course these choices cannot be made freely, because we need to use particular wavelengths to make specific observations, and telescopes cannot be made arbitrarily bigger. To give an example of resolution, an optical telescope with a primary mirror diameter of 1 m has an intrinsic resolution in the middle of the visible waveband, at 500 nm wavelength, of just over a tenth of an arcsecond on the sky. In practice, without special adaptive optics, a ground-based telescope cannot achieve its full resolution, because of atmospheric turbulence. A good mountain-top observatory can produce images with optimum resolution around half an arcsecond. For comparison the unaided human eye has a resolution for visible light of some 17 arcseconds. The Hubble space telescope produces images at this wavelength with a resolution of 0.05 arcseconds, with

Fig. 2.8 FAST China's 500 m diameter fixed radiotelescope, now with the biggest collecting area of any single dish telescope. Credit Liu Xu, Xinhua China News Agency

its 2.4 m primary mirror, free of atmospheric perturbation. Now we can make a proportional calculation to see how a radiotelescope resolves. We will consider the Arecibo diameter, 305 m, and the wavelength emitted by neutral atomic hydrogen, 21 cm. The resolution will be improved by a factor of 305 compared to the optical 1 m telescope, but will be worsened by the ratio of 21 cm to 500 nm which is 420,000. The net result will be a resolution lower by a factor 420,000/305 which is 1377. So Arecibo will have an angular resolution of 1377 × 0.1 arcseconds, which is 2.3 arcminutes. This means that radioastronomers would be at a disadvantage of over a factor 100 if they were mapping objects in atomic hydrogen compared to astronomers using visible light, even taking into account the atmospheric blurring for the optical telescopes. Another way of looking at this is that to achieve the practical resolution of an optical telescope Arecibo would need to be 100 times bigger, in other words a 30 km dish would be needed. This is a tall order even for a fixed dish. To build FAST the Chinese needed to clear a valley, and relocate 10,000 people. However, physicists came up with a solution to the problem of resolution without needing extremes in engineering. This was the introduction of radio interferometers.

A radio interferometer consists of two or more radio dishes which observe an astronomical source simultaneously, and whose signals can be combined to give a single merged signal. The easiest way to think of an interferometer is to imagine the 30 km dish mentioned above, in which most of the dish was removed, leaving a set of tens of much smaller dishes, perhaps each one of 30 m diameter, dotted over the surface. We can still move this purely imaginary dish, so that it can be pointed at an astronomical object. If only one 10,000th of the original area was left the signal would be 10,000 times smaller than that of our original 30 km diameter telescope but the resolution would not be reduced, and we could map objects using the 21 cm atomic hydrogen line with the same resolution as we have for a 1 m optical telescope. This imaginary telescope is not, however, in any way practical in engineering terms. To make it practical we separate out our individual 30 m dishes, there would be 100 of them, and we put them separately on the ground. To make our observations we point them simultaneously at the source. In order to make a good map we need to combine the signals from all of them in such a way that the original structure of the source can be reproduced. This is the process of interferometry, which needs to combine the signals from all of the 100 telescopes keeping the exact timing of arrival at each telescope. A good example of this is the Very Large Array (VLA), one of the world's major radio interferometers, situated near Socorro, New Mexico, USA. This array, built in the

1970s, and later renamed for Karl Jansky comprises 27 radio dishes each of 25 m diameter, in a Y-shaped configuration with arms 21 km long, so it has some comparable characteristics to our imaginary interferometer. A picture of the VLA is shown in Fig. 2.9.

The individual antennae can be moved along rail tracks in order to change the distances between them, rearranging the array. These changes affect the resolution of the whole instrument, so that the widest spacing gives the highest resolution, but the closest spacing gives the most completely filled array. The surfaces of the dishes and the radio receivers have been continually upgraded over the years, reaching higher sensitivities, and higher frequencies (i.e. shorter wavelengths) so that the current highest resolution possible is as high as 0.04 arcseconds at the shortest wavelengths. Radioastronomy, using interferometric techniques, now routinely achieves higher resolution on the sky than optical astronomy.

Fig. 2.9 The VLA, the Very Large Array (now the Karl G. Jansky Very Large Array) an interferometric radiotelescope at Socorro, New Mexico, USA. It comprises 27 parabolic antennae, each 25 m in diameter, whose signals combine to give the radio image of the object observed. The dishes are on rail tracks, and their separation can be varied so that the maximum "diameter" of the array can range between close to 1 and 35 km. Credit: NRAO-NSF

2.2.1 Early Interferometers and Their Results

We jumped forward in time to have a look at how a modern interferometer solves the problem of resolution for radioastronomers. Now we can take a glance at the pioneers of interferometry. The first interferometric observations were made by the Australians Joseph Pawsey, Ruby Payne-Scott and Lindsay McCready in 1946. They observed the Sun with their simple radiotelescope, just after sunrise, looking over the sea, which acted as a radio mirror. This provided a virtual extra telescope, and the resulting combined interference signal allowed them to obtain high angular resolution. They could identify a strong signal from a source much smaller than the solar disc, which they identified as a sunspot group. This method of using reflected radio waves from the sea is of limited scope, but the first real interferometer, developed by Martin Ryle and Derek Vonberg at Cambridge, England (their article reporting this was published in 1946) was the basis of future work which forms a major part of modern radioastronomy. Their technique, designed to let a number of smaller dishes act as if it were a single larger dish, was termed aperture synthesis. With this, and its modifications during the next 5 years the Cambridge group began to discover radio emission (not at 21 cm, not from atomic hydrogen) from bright sources over the sky.

Among the brightest of these was Cygnus A (see Fig. 2.10) first discovered by Reber, and whose position they later narrowed down to the point that it could be identified as a galaxy, and on the night they first observed Cygnus A they also first detected Cassiopeia A (Cass A) another extremely bright radio source, which was later identified as the expanding gaseous remnant of a supernova explosion (see Fig. 2.11).

In the following decade the Cambridge group under Ryle took a leading role in developing and using radio interferometers and detected hundreds of radio sources on the sky, publishing a series of catalogues, the 1C catalogue of some 50 sources in 1950, the 2C catalogue of 1936 sources, published in 1955 with John Shakeshaft as first author, and the 3C catalogue, also by Shakeshaft and co-authors, which observed the sources at another wavelength and at greater sensitivity. The 3C catalogue was revised and updated several times up to 1983. The radio sources detected were thought in the early 1950s to be mostly the hot outer atmospheres of stars, and were often simply referred to as "radio stars". But as the measurements of intensity at different wavelengths and of the positions of the sources improved, it became clear that essentially all the sources were objects which were not stars at all. Some were external galaxies, (such as Cygnus A) and others were supernova remnants, the

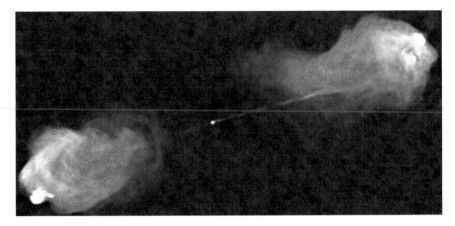

Fig. 2.10 The radio galaxy Cygnus A, imaged at a frequency 5 GHz (~6 cm wavelength) with the VLA. What we see are two huge opposing "radio lobes" of hot gas, which have been expelled almost symmetrically from the black hole at the centre of the active galaxy corresponding to Cygnus A. The zone around the black hole is the bright dot at the centre of the image. The rest of the galaxy is not picked up at this wavelength. The narrow powerful opposed jets emanating from the black hole region cause the radio lobes as they impact on the intergalactic gas. Colour is false and scaled to surface brightness. We will see Cygnus A again at other wavelengths. Credits: C. Carilli, P. Barthel, 1996, Astronomy and Astrophysics Review, Vol. 7. P.1/VLA/NRAO-NSF

gaseous expanding remains of powerful stellar explosions (such as Cassiopaeia A).

It is important to realise that these measurements were not of spectral lines, such as the 21 cm line emitted by interstellar hydrogen, but of continuous radiation, the radio equivalent of light from a filament lamp or even of a stellar atmosphere. But even this analogue is not really satisfactory. Hot filaments, hot bodies in general, including stars, emit "thermal" radiation, radiation at all wavelengths whose peak intensity depends only on the temperature of the emitting body. Our bodies emit the peak of their radiation in the infrared at around 9 μm wavelength, and the Sun's surface which is some 18 times higher in temperature emits its peak radiation at 0.5 μm, just in the middle of the visible spectrum (this is no coincidence and can be well explained by evolutionary biology). Thermal radiation is emitted with a characteristic energy distribution with wavelength (see Fig. 2.11) known as the Planck distribution, after the Nobel Laureate Max Planck who first formulated it. Stars emit most of their radiation following this distribution quite closely. In general, if the continuous spectrum of any source which is likely to have a single temperature can be measured at a minimum of three wavelengths it is possible to

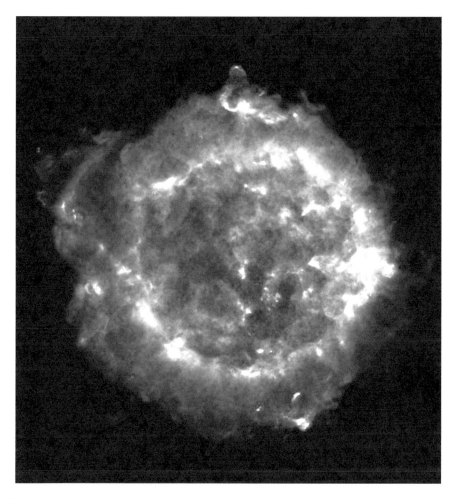

Fig. 2.11 Radio image of the supernova remnant Cassiopaeia A, taken with the Karl Jansky VLA at 11 GHz frequency (2.7 cm wavelength). This is one of the brightest radio sources on the sky and one of the first to be discovered. Credits: NRAO/AUI/NSF/L. Rudnick et al

estimate that temperature and this usually gives a fair approximation for many types of astronomical sources.

2.3   Synchrotron Radiation

As soon as radioastronomers were able to measure the intensities of their sources at several wavelengths it became clear that they did not have thermal spectra, their radio emission in no way followed the Planck distribution. Physicists looked for processes which could produce this different type of energy distribution and it became clear that one of the most convincing explanations was that of synchrotron radiation. A synchrotron is a particular type of particle accelerator which was invented by the Soviet physicist Vladimir Veksler in 1944, and first constructed by the American Edwin McMillan in 1945. Particles are accelerated in an electric field, and kept moving in a circular path with a magnetic field, which must be constantly increased as the particle moves faster order to sustain its circular motion. The synchrotron has been a fundamental instrument in particle physics, but all we will say here is that the Large Hadron Collider at CERN, Geneva, with its circular path of 27 km circumference, the leading instrument for modern particle physics research, uses the synchrotron mechanism to accelerate its protons. To maintain a particle on a circular path requires continuous acceleration, but when a charged particle is accelerated it always emits radiation. The radiation emitted by electrons in an electron synchrotron is called synchrotron radiation, and its spectrum is well known from both theory and laboratory experiment. Synchrotron light sources are very powerful, concentrated into a narrow beam, and can produce light over a wide range of the electromagnetic spectrum. They are now widely used in physics and medicine. It became clear to the radioastronomers that the type of spectra they were observing from many sources was that of synchrotron radiation. This meant that it was being produced by particles (most commonly electrons) with very high energies, moving on helical paths in magnetic fields. So here was a new tool for exploring the physics of the interstellar medium in the universe as a whole, and notably its magnetic fields. In fact, when the radioastronomers were beginning to realise that many of their newly discovered sources could be synchrotron sources, Geoffrey Burbidge interpreted the visible radiation observed by Walter Baade (both at the Mt. Wilson and Palomar observatory in the US) from a jet being emitted from the centre of the galaxy M87 as synchrotron emission, because it was polarised. He used it to estimate the energies of the particles and magnetic fields in the jet, and this type of analysis has been carried out on many astronomical sources observed at radio wavelengths since then. We now know that the high energies needed are characteristic of phenomena around the supermassive black holes in the centres of many galaxies.

2.3.1   The Crab Nebula

The Crab Nebula, (Fig. 2.12) also known by its technical names M1, NGC 1952 and Taurus A, is an expanding cloud of hot gas, the remains of a supernova explosion reported by Chinese astronomers in 1054.

It was given its popular name by William Parsons (the man who built the giant optical telescope we saw in Chap. 1) who observed it in 1840, and made

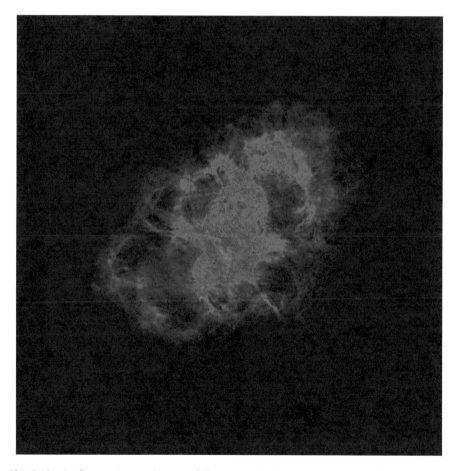

Fig. 2.12   Radio continuum image of the Crab nebula, showing synchrotron emission, taken with the Karl Jansky Very Large Array. Credits: NRAO/AUI/NSF

a drawing which looks rather like a crab. It was the first astronomical object observed and identified as the product of a supernova. It lies in the direction of the constellation of Taurus, and the radioastronomers labelled it Taurus A, one of the strongest sources of radio continuum in the sky. In 1953 the Soviet theoretical physicist Yosif Shklovsky proposed that the visible light is caused by synchrotron radiation, and this was confirmed in 1965 when it was found to be strongly polarised. But synchrotron radiation needs a strong magnetic field, and at that time it was not known how such a field could exist in a supernova remnant. As we will see shortly, the magnetic field is produced by the star left behind after the supernova exploded, a result discovered by radioastronomers in the 1960s. The radio emission from the Crab nebula gave the lead to astronomers studying supernova remnants. As supernovae produce most of the chemical elements in the universe, and as these are distributed around galaxies via the rapidly expanding gas in the remnants, finding different ways of studying them, finding "multimessengers" is of great importance, and we will see how supernova remnants crop up in many of the chapters of this book.

2.4   Radio Galaxies

Cygnus A is one of the brightest radio sources in the sky. When it was first observed and identified with an external galaxy, most astronomers thought that this galaxy was really two galaxies merging, or in more dramatic terms "colliding" and it became known as "the colliding galaxies in Cygnus". In 1953 Roger Jennison and M K das Gupta, at Jodrell Bank, showed that it was a double source of radio emission, and as observations became more sensitive deeper and deeper maps of the source showed amazing structure, with two extended jets emerging in opposite directions from the centre of a single galaxy. These jets are megaparsecs long, orders of magnitude longer than the visible size of the galaxy and end in huge clouds of gas which emit synchrotron radiation, as shown in Fig. 2.10 As observations have steadily accumulated we now know that this object is a typical radio galaxy, and that the mechanism producing the vast outflow of energy is the transformation of gravitational energy to electromagnetic energy in the region surrounding a supermassive black hole at the nucleus of the galaxy. There is no doubt that radio observations gave the chief clues to this line of research and discovery. And the most dramatic step was taken with the finding of the first quasar.

2.4.1 The Discovery of Quasars

When radioastronomers in the 1960s set out to study the range of sources found and catalogued, notably in the 3C catalogue from Cambridge, they distinguished between extended sources and point sources. The extended sources were big enough so that their structure could be mapped and the point sources were those too small to be resolved by the radio telescopes, even the interferometers. For both types of sources one of the first steps was to try to identify an optical source at the same position as the radio source, but for the point sources this was not easy, because the resolution on the sky was typically of order arcminutes, and in a circle of this diameter a typical optical image has hundreds, even thousands of different sources, mostly stars in the galaxy, with a sprinkling of external galaxies. There is a way of determining the position of a point radio source to a much greater accuracy than is allowed by the size of a single dish or even the separation of interferometer dishes. That is to observe the exact time when the source is occulted by the edge (the "limb") of the Moon, and if possible the exact time when it re-emerges. Of course this is possible only for those sources which lie along the Moon's path in the sky. In 1962 the Australian radioastronomers at the Parkes 64 m radio-telescope led by John Bolton set out to observe the point source 3C273 during a lunar occultation which occurred on August 6th. They in fact found two sources, a point source, whose position was determined to better than 10 arcseconds, and a nearby diffuse source. They sent their results to Maarten Schmidt, working with the 200 in. telescope at Mount Palomar in California, who took a spectrum of a 13th magnitude star close to their measured position. He expected that the diffuse source was emitting the radio waves, not the star, but to his surprise the spectrum showed something very remarkable. The "star" had a redshift of 0.158. Measuring the wavelengths of the very well known lines due to hydrogen, he could find their wavelengths and compare them to the wavelengths of the same lines seen in a laboratory. The difference is a measure of the velocity of the object observed. Distant objects in general are moving away from us in the expanding universe, which shifts their spectral lines towards longer wavelengths. This is known as the redshift, and it gives the velocity of the object which is moving away. For objects moving at less than some 20% of the velocity of light, c, their velocity can be calculated quite well by multiplying the redshift by c. So a redshift of 0.158 means an object that is receding with a velocity of 47,000 km/s. This value immediately tells us two things; firstly that the object cannot be within the galaxy, but must be an external much more distant object, and secondly we can estimate its distance. That is done by using the Hubble-Lemaître Law of expansion of the universe on large scales, which says that the velocity of recession of an external galaxy is proportional to its distance from us. Observations over the years which we will see in other chapters have given a value for the relation between expansion velocity and distance, the "Hubble-Lemaître Constant" as close to 70 km/s/Megaparsec. Using this the distance of 3C273 is found to be 677 Mpc. For this to be so, it must be emitting hundreds of times more power at radio wavelengths than the radio galaxies known prior to 1962. As a point object in the sky it could not be distinguished from a star, but it was clearly something very different. They were first called "quasi-stellar radio sources" which was shortened to quasar by Hong-Yee Chiu, a Chinese born American astrophysicist, in 1964 and this term is now universally used. Research into quasars has become an important part of astrophysics. Before it became well established that the nuclei of a major fraction of galaxies are supermassive black holes, with masses between one million and over a thousand million times the mass of the Sun, and that these objects have the power to produce vast quantities of radiation at many wavelengths in the volume around them as objects fall towards them, many astronomers were reluctant to believe that their redshifts were really due to the expansion of the universe because this would make their radiative emissions too powerful to be credible. But this situation was cleared up by the 1990s and now it is hard to find anyone who is a quasar sceptic. Quasars have been incorporated into a wider category called "active galactic nuclei" which we will meet when considering all of the new astronomies.

2.4.2   The Discovery of Pulsars

The discovery of pulsars is very well known, thanks to the fact that the Nobel committee, in its wisdom, decided not to award the prize to Jocelyn Bell for her major part in this work. Here we will focus on the technical aspects of the discovery. Anthony Hewish had constructed a large telescope array at Cambridge in order to study the scintillations of radio sources caused by the solar wind, which was aimed at studying the interplanetary medium itself, and using the phenomenon to measure the sizes of the radio sources, and in particular to find the smallest of these which might be quasars. This was a technically simple interferometer, built along principles devised by Martin Ryle, hardly recognisable as a telescope (Fig. 2.13) because at the long wavelength being observed an organised set of wire makes a satisfactory instrument. On 28th November 1968 Jocelyn Bell, Hewish's graduate student,

Fig. 2.13 Radiotelescope at Cambridge, U.K. with which Jocelyn Bell and Anthony Hewish discovered the first pulsars. Courtesy of the Mullard Radio Astronomy Observatory, University of Cambridge

noticed a regular repeating signal, shown on a long strip chart, with peaks every 1.33 s. A short extract from one of the chart is shown in Fig. 2.14.

The signal was regular, but intermittent, and appeared to come from a single point on the sky. After apparently ruling out known natural and human sources, Bell and Hewish nicknamed the source LGM-1 (for "little green men"). By December Bell had discovered three further similar sources in different parts of the sky. These observations were published in Nature magazine in February 1968. Shortly thereafter the British astrophysicist Thomas Gold proposed that signals such as these could be emitted from a rotating neutron star, which may emit a narrow beam of synchrotron radiation that is detected every time it crosses our line of sight, just like the rotating beam from a lighthouse. A neutron star is produced when a star of intermediate mass explodes as a supernova: less massive stars produce white dwarfs, and more massive stars produce black holes. In all three cases the outer layers of the star are blown off, and the remaining core is a highly compressed star, with an enormous density. Black holes are denser than neutron stars which are denser than white dwarfs. In a neutron star the atoms are so compressed that effectively they have absorbed their own electrons, turning their protons into neutrons, so that the whole body of the star is made of neutrons. Although the physics

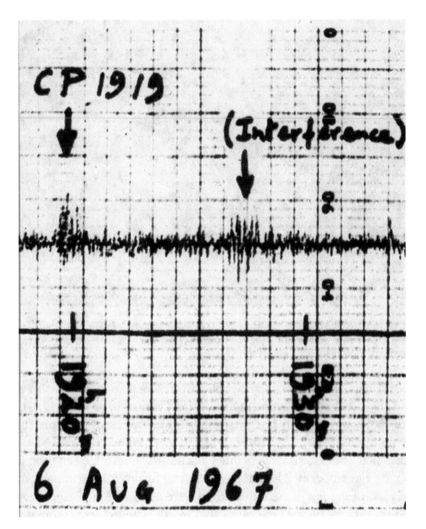

Fig. 2.14 Strip chart with a blip (a "squiggle") from CP1919 one of the first pulsars. Courtesy of the Mullard Radio Astronomy Observatory, University of Cambridge

is considerably more complex than this, and at the surface all three types of particles exist, this does tell us that the density of a neutron star is that of an atomic nucleus. To give a general idea, a teaspoon full of the matter in a neutron star would weigh a thousand million tonnes. Ryle and Hewish won the Nobel prize in 1974, Ryle essentially for his pioneering work in radiointerferometry, and Hewish for the discovery of the pulsars. This was the first time the Nobel Prize in Physics had been awarded to astronomers. The radioastronomical community contended that this was because radioastronomy was "real physics".

Astronomers began to search for the optical equivalents of pulsars. The Italian astrophysicist Franco Pacini, who in 1967 had proposed that rotating neutron stars should emit beamed synchrotron radiation suggested that the Crab nebula might be energised by such an object in its interior. Pulsating radio emission from the Crab nebula at some 30 pulses/s was first detected at the 300 m dish at Arecibo, Puerto Rico, which has since been used to discover many pulsars. This pulsar was then detected at optical wavelengths by the British astronomers John Cocke, Mike Disney and David Taylor working at the Steward Observatory in Arizona, who could find it because they knew the period of its pulsations from the previous radio data. Figure 2.15a shows the Crab nebula in a combination of visible light from the Hubble Space Telescope, and an X-ray image from the Chandra satellite which we will meet in the chapter on X-ray astronomy. This image clearly gives a dynamic impression of a rotating source in the centre, and is notably different from the radio continuum image we saw in Fig. 2.12, and from the optical emission line image in Fig. 2.15b. These very different illustrations show that the more messengers we have for a given object the more we can expect to learn about it.

2.4.3   Pulsars and General Relativity

In 1974 Russell Hulse, a graduate student at the University of Massachusetts, was observing a pulsar at the 305 m Arecibo telescope when he noticed that the period of its pulses, the time between a pair of pulses, varied regularly. He and his supervisor Joseph Taylor deduced that this was because the neutron star emitting the pulses is in orbit around another star. When the pulsar is approaching us the interval between the pulses is shortened, and when it is moving away the interval is lengthened. They realised quickly that the behaviour of this pulsar and in particular the short period of rotation about its companion, 7.75 h, meant that the two stars were quite close and that orbiting in a strong gravitational field the pulsar ought to show measurable effects due to general relativity. The complete theoretical description of the pulsar's behaviour is too complex to describe here, but we can mention two of the effects. The first is similar to the effect which makes the nearest point of Mercury's orbit to the sun change position, and which was used to show that General Relativity gives a better description of gravity than Newton's theory. For Mercury the change in the orbit's orientation close to the Sun is 43 arcseconds per century, but for the Hulse-Taylor pulsar it is 4.2° per year! The

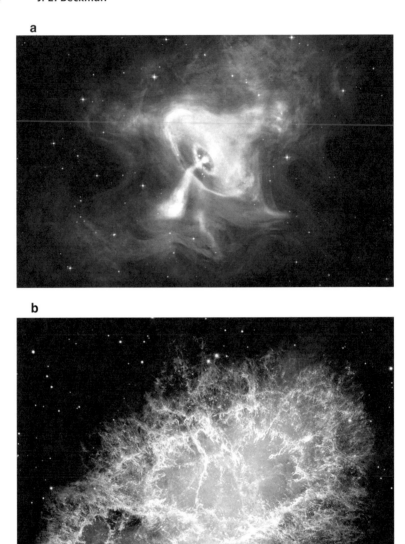

Fig. 2.15 (a) The Crab Nebula pulsar. Image in optical light from the Hubble Space Telescope combined with X-ray emission from the Chandra satellite. The pulsar is at the centre of the spinning material. Credits: NASA/ESA/HST/CXC/J. Hester et al. (b) Mosaic image by the Hubble Space Telescope of the Crab Nebula. The blue haze around the

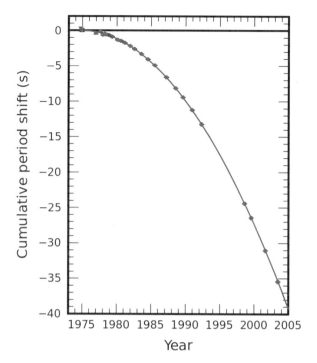

Fig. 2.16 Cumulative change in the time of periastron of the binary pulsar PSR B1913+16 as it orbits its companion. The points are the observations, showing the small error bars, and the line is the prediction of General Relativity for this system. Credits: J.Weisberg, D.J. Nice, J.H. Taylor, 2010, Astrophysical Journal, Vol 722, p. 1030

second effect is the change in the 7.75 h rotation period, predicted by General Relativity because the neutron star (the pulsar) should be emitting gravitational waves, which will make its period get shorter as it spirals in towards its companion. All of the information needed to test these predictions is to be found by careful timing of the pulses over a sufficient period of time. It is now possible to time the arrival of any pulse to within 15 ms. In Fig. 2.16 you can see the measured shortening of the period of the Hulse-Taylor pulsar plotted against time, together with the theoretical prediction of the effect of gravitational radiation from General Relativity. The agreement is spectacularly good.

In a very real sense these observations were a proof of the existence of gravitational waves before their direct detection in 2015 by LIGO. The discovery of the binary pulsar was the first in a set further similar objects, including a double neutron star where both members are pulsars, by the Italian astronomer Marta Burgay and coworkers in 2003. This latter binary

---

Fig. 2.15 (continued) centre is synchrotron radiation from the pulsar, and in the outer filaments blue light comes from neutral oxygen, green from singly ionised sulphur and red from doubly ionised oxygen. Credit: NASA/ESA/HST/J. Hester/A. Loll (ASU)

pair is being used in five separate ways to test General Relativity against rival theories. The importance for physics of the Hulse-Taylor pulsar was recognised in 1993 when they shared the Nobel Prize.

2.5 CO and the Three Phases of the Interstellar Medium

The detection of atomic hydrogen in the interstellar medium of our Galaxy and of other galaxies using the 21 cm line allowed astronomers to map complete galaxies and measure their internal dynamics. As hydrogen is the fundamental constituent of stars, measurements of atomic hydrogen, HI, provided a basis for the studies of star formation which took off in the 1950s and 1960s and were given an impulse when infrared astronomy let us look through the dust into star forming clouds. But physicists knew that the cool conditions in many zones of interstellar space must be favourable for the formation of molecular hydrogen, H_2, formed when two atoms combine to form a stable molecule. The problem is that H_2 is a symmetric molecule, and the rules of quantum mechanics do not allow symmetric molecules to emit radiation (or only very weak radiation) so there is no equivalent of 21 cm radiation for H_2. This meant that a major component of all galaxies could not be directly observed. This situation was changed in 1970 when Keith Jefferts, Sandy Weinreb and Robert Wilson detected the carbon monoxide, CO, molecule within the Orion Nebula. They used a receiver designed at the American company Bell Labs, which at its peak from 1960 to 1980 showed that a private company could make fundamental contributions to physics at the highest level. We will meet Robert Wilson again in the context of the discovery of the Cosmic Microwave Background radiation. The CO detection was made at a wavelength of 2.7 mm, a frequency of 115.8 GHz which was at a higher frequency than those current at the time in microwave receiver technology. The demands of this new branch of radioastronomy, millimetre and submillimetre wave spectroscopy, pushed all the technologies in this field, giving an example of how astronomy can be at the forefront of achievement in physics. The detection of CO was very important because it does have emission lines which can be produced at the low temperatures in interstellar space, but it needs to be in a cloud with a certain minimum density before it emits these lines. This density is supplied by molecular hydrogen, so that a cloud with a density in H_2 of more than 1000 molecules per cubic centimetre will emit CO lines. Thus measuring and mapping CO gave astronomers the tool they needed to map the molecular hydrogen component of galaxies. In addition the conditions within the molecular clouds are those needed for star formation, so CO gave astronomers a powerful tool to explore the star forming process.

The interstellar medium in galaxies can be well characterised by the presence of one of the three main phases of hydrogen. The equilibrium between the phases depends on the temperature and the density of each specific cloud, and also on the radiation field in the given part of the galaxy. We can specify the phases as atomic hydrogen, HI, molecular hydrogen H_2, and ionised hydrogen HII, which is atomic hydrogen which has lost its electron. Astronomers need to combine information about the three phases in order to understand the interstellar medium and its key role in star formation. Radioastronomy gives us the information we need about the atomic and molecular phases, and is therefore responsible for a major slice of the progress made in this field. Ionised hydrogen is found mainly in the gas around hot young stars, although a smaller fraction of interstellar hydrogen is ionised almost everywhere in a galaxy. It is somewhat ironic that although one of the best ways to observe it is to use the H-alpha line, emitted in the red part of the visible spectrum, which should be open to conventional visible spectroscopy, progress in measuring HII has been far more limited than that in HI and H_2, but recent work is improving this situation, as we have seen in the chapter on optical astronomy.

2.5.1 Molecular Radioastronomy

Although CO has been the fundamental molecule observed in the interstellar medium, the technical ability to extend the radio spectrum to ever shorter wavelengths, into the millimetre and submillimetre range until it meets the far infrared, has led to the discovery and measurement of numerous molecules in interstellar space, some in the coolest depths of space, others in the warmer regions close to stars. At the present time over 200 different molecules have been identified ranging from the simplest, such as CO and sulphur monoxide SO, to molecules with 11 atoms such as acetic acid CH_3COOH and amino-acetonitrile NH_2CH_2CN. Of course scientists are on the lookout for the molecules which form the basis of life, amino acids or DNA for example, but the more complex the molecule the smaller the quantities, and the weaker the signal will be. A molecule which has been found in quantity, and which naturally aroused the curiosity of the media, is ethyl alcohol, C_2H_5OH. In 1995 a specific molecular cloud in the direction of the constellation of Aquila was measured to contain the same quantity of ethyl alcohol as 400 million million million million pints of beer! Molecular astronomy has blossomed in the 50

Fig. 2.17 View of the ALMA (Atacama Large Millimetre Array) millimetre and submillimetre wave telescope , on the Chajnantor plateau, in the Atacama desert, Chile at 5000 m above sea level. It comprises 66 antennae of 12 and 7 m diameter. Credits: ESO/NSF-NRAO/NRCC/NAOJ/ASIAA

years since CO was first detected, and would need a book on its own to describe its findings. It has spawned specialised instrumentation, and it is worth showing its outstanding telescopic installation: ALMA. The Atacama Large Millimetre/submillimetre Array. Figure 2.17 is an interferometer comprising 66 dishes each of 12 m in diameter, which operates between 9.6 and 0.3 mm in wavelength, (31–1000 GHz in frequency). It is a collaboration between European, North American and East Asian astronomers, and is currently the most expensive ground-based telescope.

It was built at 5000 m altitude in one of the driest places on Earth in order to minimise the quantity of water vapour in the air above the telescopes, as water vapour absorption limits the penetration of submillimetre wave radiation through the atmosphere. As well as being the most powerful telescope for molecular astronomy ALMA has the best angular resolution of any single ground-based telescope, some 6 milliarcseconds, and can be used to obtain images with astonishing detail in its chosen wavebands. Figure 2.18 shows an ALMA image of a young star HL Tauri, surrounded by a disc of dusty gas, which has been broken into separate rings by newly forming planets within the disc. It is an illustration of the contribution of ALMA in the burgeoning field of planet formation. The image uses the full angular resolution with the

Fig. 2.18 ALMA image of the dusty protoplanetary cloud around the young star HL Tauri. The disc shape, and the division of the cloud into separate rings is part of the process of planet formation, similar to the production of the solar system. Credits: C.L.Brogan, L.M. Pérez, /the ALMA Partnership, 2016, Astrophysical Journal Letters, Vol. 808, p. L3

maximum distance of 16 km between the most widely separated pair of antennas.

2.6 The Cosmic Microwave Background

It has been said that the most significant contribution of radioastronomy to the whole of science was the discovery of the cosmic microwave background (CMB) radiation. This story has been told many times, and we will not repeat it here, except to give a picture (Fig. 2.18) of the radio horn telescope with which Arno Penzias and Robert Wilson first measured the radiation in 1964, (published in 1965), for which they were awarded the Nobel Prize in 1978 (Fig. 2.19).

Fig. 2.19 Arno Penzias (wearing glasses) and Robert Wilson with the radio antenna at Bell Labs. Homdel New Jersey (USA) with which they discovered the cosmic microwave background radiation. Wilson also participated in the first measurement of the CO molecule in the interstellar medium. Credit:Bell Labs/Nokia

Taken together with the expansion of the universe measured by the expansion of its galaxy systems on a large scale, this radiation, which fills the sky, is taken as the main pillar of observational support to the Big Bang theory. The clinching set of observations of the CMB was made by measurements from satellites, notably by NASA's Cosmic Background Explorer (COBE). Between 1989 and 1992 COBE made two outstanding measurements: the spectrum of the CMB with the FIRAS spectrophotometer, and the anisotropy with the DMR, the Differential Microwave Radiometer. We will discuss these experiments and their results in the chapter on cosmology and particle physics.

2.7   Very Long Baseline Interferometry

We have seen how interferometry has enabled radioastronomers to equal and then exceed the angular resolution of optical telescopes, using two or more separate antennae and combining the signals. There is no limit, in principle, to the separation of the antennae, although in practice the differences in atmospheric transmission above the telescopes must be taken very carefully into account. This freedom encouraged radioastronomers to place their antennae at ever increasing distances to obtain higher and higher angular resolution. The first time this separation was tried over long distances was in Canada in 1967, using telescopes at the Dominion Astrophysical Observatory in British Columbia and at the Algonquin Observatory in Ontario, a separation of 3074 km, which marked the beginning of VLBI (very long baseline interferometry). Following this, an array of widely separated dishes was built across the United States, the Very Long Baseline Array (VLBA). Now there are VLBI systems working across all the continents, between continents, and even between ground and space. A milestone was achieved in 1976 when radiotelescopes in the US, the USSR, and Australia were linked to achieve microarsecond resolution, in order to measure the emission from OH masers in interstellar space. This type of measurements allows astronomers to measure movements on the sky across the line of sight, angular movements which are too small to estimate in any other way, and complements the normal and much easier technique of measuring velocities along the line of sight by the Doppler shift of spectral lines, which we saw in the chapter on optical astronomy, but which works at all wavelengths. Originally VLBI records had to be taken separately at each observatory, and brought together physically by transporting magnetic tapes or discs, but since 2011 telescopes have been linked in real time using Gbit per second radio links, a technique termed e-VLBI. Also in 2011 Russia launched a VLBI satellite RadioAstron, which orbits at 390,000 km from the ground and provides the longest baseline, highest resolution data of any system until now, below 10 microarcseconds. This is the equivalent of resolving the width of a human hair at a distance of 20,000 km. VLBI is so powerful that it is now routinely used for a variety of geophysical observations, such as measurements of the speed and direction of the drift of the Earth's tectonic plates, the deformation of land masses, variation in the Earth's orientation, and the length of the day. It could be used as an alternative to satellite GPS for navigation, using pulsar signals as frames of reference. A spectacular achievement of VLBI, and at millimetre wavelengths, was the "photograph" of the immediate envelope of the black hole at the centre of the galaxy M87 shown in Fig. 2.20. In Fig., 2.20(a) we see the sites of the millimetre wave telescopes around the world which combined to make the measurement, collectively termed the Event Horizon Telescope, and in Fig. 2.20b we show an image taken with the telescope. The southern part of the imaged annulus is brighter than the northern part, showing that the radiation comes from a component which is moving towards us as it spirals into the black hole.

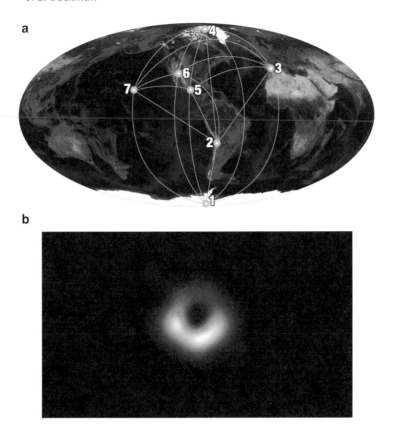

Fig. 2.20 (a) Sites of the telescopes making up the Event Horizon Telescope, a millimetre wave VLBI instrument, with angular resolution sufficiently high that it could image the immediate surroundiof the supermassive black hole at the centre of the elliptical galaxy M87. The participating telescopes in 2017 are 1. The South Pole Telescope (SPT), 2. ALMA and also APEX (Chile); 3. The IRAM 30 m telescope (FRANCE), 4. The Greenland telescope 5. The Gran Telescopio Milimétrico Alfonso Serrano, GMT (México), 6. CARMA, (California, USA), 7. James Clerk Maxwell Telescope, and the Submillimeter Array, SMA, Hawaii, USA. (b) The image of the shadow of the black hole at the centre of M87 surrounded by its atmosphere of matter falling into the black hole and being excited to emit radiation. The diameter of the outer edge of the illuminated annulus is some 60 μarcsec (microarcseconds). Credits: Event Horizon Telescope consortium

2.8   The Square Kilometre Array

As explained when they were first mentioned, interferometers were designed to maximise the resolution of radiotelescopes. But the total collecting area of an interferometer is much less, usually of order 100 times less, than the area of a filled dish with the same equivalent diameter. The Square Kilometre Array (SKA) project aims to overcome this by covering, as its name implies, a square kilometre of surface with a major array of telescopes which are all cross-linked to form a giant interferometric system. The project is very complex but the world's radioastronomers joined forces to promote it, and it is an approved joint project shared among 11 countries: Australia, Canada, China, India, Italy, New Zealand, South Africa, Spain, Sweden, The Netherlands, United Kingdom. A project to develop new technology for the dishes and the electronics is assigned to a consortium of universities led by Cornell, in the US. After considering rival site proposals, the SKA governing body decided to split the telescope and place it on two sites, one in Western Australia and the other in South Africa, with an administrative and technical headquarters at Jodrell Bank in the UK. When complete the telescope will comprise several thousand small antennae with a variety of electronic links between them, which will enable the telescope to observe several fields on the sky at once. The full frequency range, eventually from 50 MHz to 14 GHz, with a later extension to 30 GHz, requires the SKA to be composed of several independent segments: a low frequency segment, from 50 to 350 MHz, and a mid-frequency segment, from 350 MHz to 14 GHz. Several different types of antennae will be used, including simple radio dipoles, and a few thousand 12 m dishes. One of the biggest challenges of the SKA is the huge volume of data it will collect, one estimate being the equivalent of 35,000 DVD's of data every day, or "a whole world wide web per day". The project will push capacity for data reduction and storage well beyond its present limits, and is a good example of pure science stimulating new technology.

To prepare for the SKA there are several "pathfinder" and "precursor" telescopes already working. The pathfinders are existing telescopes which are being constantly enhanced in their capabilities, and include the Expanded Vary Large Array in the US. The precursors are arrays built or under construction at the SKA sites. They include ASKAP, comprising 36 12 m dishes, at the Australian site, using and testing advanced technologies such as phased array feeds which provide a very wide field of view. At the South African site MeerKat, an array of 64 dishes each of 13.5 m diameter is under development and has recently been put through a commissioning run. To show its paces it

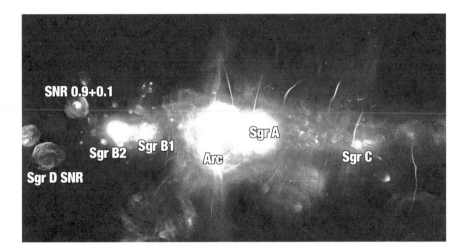

Fig. 2.21 The centre of the Galaxy imaged by the MeerKAT telescope, one of the SKA pathfinders. The image was made at 23 cm wavelength and the bubbles and filaments are emitting synchrotron radiation from stellar explosions in the crowded central region. Dense dust prevents imaging this region at optical wavelengths, and even in the near infrared. Well-known individual radio sources are named. Credits: MeerKAT/SAAO

took a map of the area around the centre of the Milky Way with the spectacular result shown in Fig. 2.21.

There is no doubt that radioastronomy was the pioneer of the new astronomies and has made contributions over a wide range of fields, most notably in extragalactic astrophysics and cosmology, but also in the field of star formation. With ALMA and FAST in full performance and with the SKA under development, radioastronomy's future will be at least as interesting as its past.

Further Reading

Book. Burke, Bernard S., Graham-Smith, Francis, Wilkinson, Peter F. An Introduction to Radio Astronomy. Cambridge. 4th Edition 2019

Electronic book: https://en.wikibooks.org/wiki/Category:Book:Pulsars_and_neutron_stars

Free book Chapter. Arnold, S. 2014. The History of Radio Astronomy, https://doi.org/10.1007/978-1-4614-8157-7_1. This is a single chapter in a book, Getting Started in Radio Astronomy. The Patrick Moore Practical Astronomy Series. Springer, New York. The whole book is available in print ISBN: 978-1-4614-8156-0 or online. ISBN: 978-1-4614-8157-7

Web reference to National Radioastronomy Observatory on-line article: https://public.nrao.edu/radio-astronomy/the-history-of-radio-astronomy/

Book: Fielding, John (2006), Amateur Radio Astronomy (Potters Bar: Radio Society of Great Britain) ISBN 1-905086-16-4. Good for bright students with some knowledge of radio components and circuits

3

Infrared Astronomy

3.1 Pioneers of Infrared Astronomy

Infrared astronomy started seriously in the mid-twentieth century, but a few much earlier landmarks are worth mentioning. In 1800 William Herschel noted that when he explored the energy in the spectrum of sunlight using a thermometer, the maximum response was obtained when the thermometer was just off the edge of the visible light, at the red end of the spectrum. The next step, the detection of infrared radiation from the Moon was by the Astronomer Royal for Scotland, Charles Piazzi Smyth, during an expedition to the mountains of Tenerife. This was made with a 1.88 m equatorial telescope, lugged up from sea level to Mount Guajara at 2710 m above sea level on a donkey. He used a thermopile which, just as for Herschel before him, detected more "heat" beyond the red end of the visible spectrum than in the visible part itself. This remarkable feat was crowned by taking the telescope to 3500 m and showing that the infrared was stronger relative to the visible radiation when observing from that height. This was the first serious attempt to prove that astronomy from mountain tops gives advantages. One result in the long term was that the two professional observatories in the Canary Islands, one on Tenerife, the other on La Palma, are now among the leading sites in the world for astronomy. Piazzi Smyth calibrated the Moon's infrared by comparing it with the heat from a candle, but of prescient importance was his technique for separating the Moon's emission from that of the rest of the sky. He did this by "chopping", using a moving mirror to switch rapidly, first pointing at the sky and then at the Moon, before taking the difference. This is still the method used for ground based infrared observations today. In the first

© Springer Nature Switzerland AG 2021
J. E. Beckman, *Multimessenger Astronomy*, Astronomers' Universe,
https://doi.org/10.1007/978-3-030-68372-6_3

decade of the twentieth century Ernest Nichols made detections of two very bright stars, Arcturus and Vega, in the infrared, and although he himself was not convinced by the results, the flux he recorded agrees with the modern value. In the 1920s Seth Nicholson and Edison Petit developed thermocouples which were infrared sensitive. They estimated the temperature of the Moon and detected planets and a couple of hundred stars, as well as observing sunspots. Their estimates of the temperatures of nearby giant stars led to the first estimates of stellar radii: by combining measured values of temperature, distance, and total radiative power the radius of a star can be found, assuming as an approximation that it radiates as a black body at the measured temperature. In the 1940s and 1950s Peter Fellgett and Albert Whitford used lead sulphide detector cells to measure stellar temperatures, and Whitford began to work on the problem of interstellar dust, which absorbs light with different efficiency at different wavelengths and can, if not well calibrated, play havoc with stellar temperature measurements; the dust is less opaque in the infrared. In the 1950s and 1960s Harold Johnson laid down the basic passbands to be used for measuring at infrared wavelengths. In a review of infrared astronomical measurements in 1966 he collated and standardised terms and definitions. He called the near infrared the range from the visible to 1 micrometre (µm), a range accessible to the photomultipliers of his day, which were developed for the visible, but extended somewhat further. He termed the intermediate infrared the range from 1 to 4 µm, a range for optimum sensitivity of lead sulphide cells, and the far infrared as ranging from 4 to 22 µm. Since then techniques have advanced, and these boundaries have lost their utility. There are no longer accepted rigid definitions of near, middle, and far infrared and in any case the practical long wavelength boundary has moved outwards to meet the radio spectrum in the submillimetre range, at a few hundred µm, i.e. a few tenths of a millimetre. So nowadays people commonly refer to the near infrared as going from the edge of the red, at around 0.7 µm, out to 5 µm wavelength. This includes the six standard passbands for photometry of stars and indeed all objects which emit a continuous spectrum: the I band centred at 0.9 µm the J band, at 1.25 µm, H, centred at 1.65 µm, K centred at 2.1 µm, L at 3.5 µm, and M at 4.8 µm. The two middle infrared bands are N between 7.5 µm and 14.5 µm, and Q between 17 and 25 µm. There is a single demarcated far infrared band, Z, from 28 to 40 µm, and the rest of a much wider far infrared band goes from 330 to 370 µm. The gaps between some of these bands are due to the intense water vapour absorption which makes observations from the ground impossible. The basic bands, their letter denominations and bandwidths were systematised by Johnson, who made infrared photometric measurements of hundreds of stars in the bands up to N.

3.2 Why Astronomers Want to Use the Infrared

The main reason why astronomers want to observe in the infrared is to observe the cool universe. To understand this we can look at Fig. 3.1, which shows the distribution of radiation by wavelength from bodies at four different temperatures. It is a simplified version of real radiation, because it corresponds to the temperatures of idealised bodies, referred to in physics as "black" bodies, but the graphs do give a generally valid picture of the way real objects, including stars and other astronomical bodies, behave.

To summarise the meaning of the graph, if two bodies of the same surface area have different temperatures, the cooler of the two emits less radiation, and also the peak of its emission is at a longer wavelength. To take an example of the wavelength difference, the Sun whose surface is at close to 5500 K emits the peak of its radiation at a wavelength close to 500 nm, (0.5 μm) which is in the green part of the visible wavelength range, and the general aspect of its

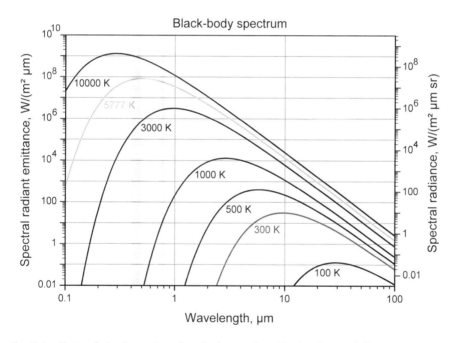

Fig. 3.1 Plots of the intensity of radiation emitted by bodies at different temperatures, as functions of wavelength in microns (μm). Temperatures are in Kelvins, K, where 300 K is 27 C (Celsius) so that for high temperatures the scales are nearly the same. The visible spectrum range is shown as the rainbow block just below 1 μm. To the left is the ultraviolet, to the right the infrared. Infrared astronomy deals mainly with cool objects which radiate between 1 μm and 300 μm wavelength. Credit: Sch, CC BY-SA 3.0 http://creativecommons.org/licenses/by-sa/3.0/, via Wikimedia Commons

light is yellow. The moon's surface ranges in temperature from 400 to 100 K depending on whether it is in full sunlight or in full darkness. The peak of its radiation varies between 5 and 25 μm, and is well into the infrared range. We know that we see the Moon in visible light by its reflection of sunlight, and not by its own emitted radiation. Only when infrared astronomers began to observe the Moon could they make direct measurements of its properties. But this is an illustrative example, not especially typical. There are many objects and processes in the universe which are cool, and which are therefore best observed in the infrared. Among the most important processes is that of star formation, which is a very active research area, and which could not be tackled without infrared techniques. Also a large family of stars less massive than the Sun emits most of its energy at infrared wavelengths, and planets, not only those of our own solar system, but the rapidly growing number of exoplanets, are observable directly only in the infrared.

A more recent and certainly powerful reason to observe in the infrared is that with larger telescopes and more sensitive detectors we can observe fainter and fainter objects which are further and further from us. This means that we can see galaxies at great distances which are moving away from us due to cosmic expansion. This expansion gives the galaxies velocities of recession which shift the light we receive from them to longer wavelengths, the famous "red shift". Nowadays we can detect galaxies which emitted their light when the universe had only a fraction of its present age. If this fraction is ½, the light is shifted to twice the wavelength at which it was emitted. This means that visible light is shifted into the near infrared. If the galaxy emitted its light when the universe was a quarter of its present age, the wavelength received is shifted to four times its original value. So as we are able to look further and further out into the universe, and further back in time we need to use first the near infrared, and then the middle infrared just to see the same kinds of images and spectra that we are accustomed to in the visible.

3.3   The Problem of Interstellar Dust

But another powerful reason to observe in the infrared is because large zones within our Galaxy and in other galaxies are obscured to us in the visible range by clouds of interstellar dust. In the Milky Way astronomers have learned from observations that the large clouds of interstellar gas, mostly hydrogen, have an admixture of dust particles which have been produced by the condensation of heavier elements and compounds produced in stars, and ejected into the interstellar medium (the ISM) either during their lifetimes, for lower mass stars, or by the explosions which terminate their lives, for higher mass stars. This dust forms no more than 1% of the ISM by mass, (this is a global figure; there are certainly wide variations in time and place within the Galaxy) and is in the form of small grains, with sizes ranging from a few nm to a few μm. Although the mass fraction is small, over distances of light-year scales the total effect of the dust can be to block the light from any object, to a greater or lesser degree. The efficiency of this blocking depends on both the number density of the dust grains and their size distribution, but one empirical key fact is that dust blocking, normally referred to as "extinction" is increasingly efficient at shorter wavelengths. This means that a star seen through a typical dust cloud will be redder than a star of the same type (same temperature) observed along a clear line of sight with no dust extinction. This is why dust extinction is often referred to as reddening.

Figure 3.2 is a photograph of the Milky Way in visible light; the multiple dark zones are not zones without stars, but gas and dust clouds, where the interstellar dust is blocking the light from the stars behind the clouds from reaching us. Figure 3.3 shows a measured "extinction curve" for the Milky Way and for different regions in two nearby galaxies, the Magellanic clouds. For historical reasons, (and in this astronomers are often eccentric, and are criticised by fellow physicists) these extinction curves are not plotted against wavelength λ but against inverse wavelength 1/λ so that the visible range in

Fig. 3.2 The Milky Way photographed over the Island of La Palma, Canary Islands, with the Gran Telescopio Canarias in the foreground. The dark lanes in the Milky Way are caused by absorption of light by interstellar dust clouds. Credits: GTC/IAC/ Daniel López

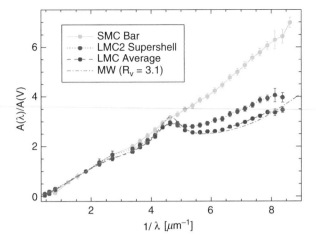

Fig. 3.3 The extinction due to interstellar dust; an average for the Milky Way (MW), an average for the Large Magellanic Cloud (LMC Average)), and for a specific zone of the Large Magellanic Cloud (LMC Supershell) and for the bar of the Small Magellanic Cloud (SMC Bar). The x-axis is in reciprocal wavelength (1/λ) so the extinction is rising towards shorter wavelengths. The A(λ) is plotted relative to the value at the visible wavelength A(V) and goes down to small values in the infrared towards the left of the graph. Credit: Karl Gordon

the figure lies between $1/\lambda = 1.3$, which is a wavelength of 750 nm or 0.75 μm, in the red, and $1/\lambda = 2.6$ which is a wavelength of 380 nm or 0.38 μm, in the violet. There are features of interest in this curve. Firstly we can see that it rises from left to right, which is in the direction of decreasing wavelength. Secondly it is quite a smooth curve, although it has a bump a $1/\lambda = 4.65$, in the near ultraviolet. This feature is certainly of interest for those studying the interstellar medium but we will not follow it up here. Thirdly in the infrared, which is the range of all values of $1/\lambda$ smaller than 1.3, the extinction is quite small. It appears to go virtually to zero at the longest wavelengths, and although a much more detailed look at this curve for values of $1/\lambda$ less than 0.1 has been obtained and studied, all we need to know here is that the extinction in the near infrared is quite small, and in the middle and far infrared is very small. As star forming regions are typically shrouded in clouds of gas and dust, this alone is sufficient reason to study astronomy in the infrared.

The interstellar dust absorbs and scatters light at optical and near infrared wavelength because it is cool, certainly much cooler than the surfaces of stars in general. However all matter radiates, and interstellar dust is no exception. From Fig. 3.1 we can see that the peak wavelength emitted by any object is essentially inversely proportional to its temperature so that interstellar dust at

100 K, which is a typical temperature of clouds of atomic hydrogen will emit most of its radiation in the far infrared at 25 μm while dust at 10 K which is a temperature quite common in dense molecular clouds which are not near stars, will emit where the infrared runs into the submillimetre range at 250 μm and longer wavelengths We will receive the equivalent emission from distant galaxies at even longer wavelengths due to the cosmological redshift

3.4 The Start of Modern Infrared Astronomy

The first big advances in infrared astronomy came in the 1960s and early 1970s with the development of infrared detectors based on increasing under-standing of solid state physics and technology. The first major survey of the sky in the near infrared was carried out by Gerry Neugebauer and Robert Leighton in 1969. It was called the Two Micron Survey, and the observations were carried out from Mount Wilson in California, which had been a classical site for optical astronomy with the 100 in. telescope, but where light pollu-tion from Los Angeles was impeding good optical astronomy by the 1960s. Infrared astronomy could still be usefully done. Among the outstanding sources observed in this survey were the galactic centre, and the Becklin-Neugebauer object in the Orion Nebula, the brightest near infrared source in the sky. It was particularly important to use infrared to explore the centre of the Galaxy because the gas and dust are strongly compressed into the Galactic plane and as we are in the plane, lines of sight to the centre pass through these major sources of extinction, so there is no hope of observing at visible wave-lengths. The most striking use of infrared in this context was reported in 2018 by a group led by Reinhard Genzel at the Max Planck Institut für Extraterrestrische Physik at Garching, near Munich. They used several tele-scopes of the European Southern Observatory in Chile to follow the orbits of 28 stars around the centre for 16 years at high precision and from the orbits were able to state categorically that the nucleus of the Galaxy must be a super-massive black hole. They also confirmed a specific prediction of General Relativity about the rate of precession of the orbit of a star around a black hole. Figure 3.4 is an infrared picture converted to visible wavelengths for viewing, of some of these stars.

Their work is continuing and they will eventually use interferometry between several of the four large telescopes of ESO's Very Large Telescope (VLT) to improve the precision of these already amazing results. The Becklin-Neugebauer object is important because the Orion nebula is the nearest star forming region to us, and this object is the brightest of a considerable set of

Fig. 3.4 Stars in the central zone of the Milky Way, very close to the central supermassive blackhole.A team led by Reinhard Genzel using high resolution imaging in the infrared (to penetrate the dust) with telescopes at the European Southern observatory measured the orbits of a group of these stars for over 16 years, deriving the mass of the black hole with unprecedented accuracy. For this result Genzel was awarde the Nobel Prize in Physics, shared with Andrea Ghez, in 2020. Credits: Max Planck Institut für Extraterrestrische Physik (MPE)/ Reinhard Genzel

young stars, and stars in formation. Most of our knowledge about the physical conditions in which stars form have been gained using the infrared.

3.5   Observational Platforms for Infrared Astronomy

Although near infrared radiation penetrates the atmosphere and reaches ground-based sites, if we look at Fig. 3.5 we can see why, as techniques improved to allow sensitive detection in the middle and far infrared, astronomers looked for new platforms for their work.

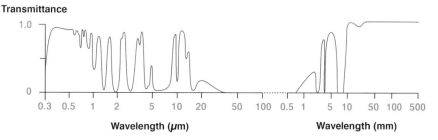

Fig. 3.5 Summary of the transmittance of the Earth's atmosphere for radiation coming from outside (astronomical sources) . The curve shows the fraction of light reaching the ground at a given wavelength, from the visible through the whole of the infrared range into the cm wavelength radio range. There is a major window in the optical range, and lesser windows in the infrared. To observe well in much of the infrared high altitude and space platforms are needed. Credit: IAC-UC3

The figure shows the transparency of the atmosphere above a typical ground-based site. In the optical and a large part of the radio range it is fully transparent, at wavelengths shorter than optical, which include the ultraviolet and X-ray ranges it is essentially opaque, while the infrared range is an intermediate case. There are some windows in wavelength which are quite transparent, and through which ground-based observations can be made, particularly from high altitude sites where the water vapour, the main source of opacity in this range, is much reduced. But parts of the infrared are fully, or almost fully blocked. Unlike the X-ray astronomers, who always knew that the only way for them to observe was using satellites, the infrared astronomers adopted a multiple approach. They used the ground, particularly high mountain sites, for the most transparent wavelengths, then they began to use aircraft for the ranges of intermediate transparency, a few experiments were made with high altitude balloons, and then came the epoch of infrared observations with satellites for the opaque wavelengths.

3.5.1 Infrared Astronomy from Aircraft

Some of the earliest work in IR astronomy from aircraft was carried out by Frank Low, at the University of Arizona, a pioneer of far infrared detectors, who in 1966 flew his detectors behind a 12 in. telescope on a small plane, a Lear Jet belonging to NASA. He discovered that the amount of infrared emission from both Jupiter and Saturn was greater than the energy they receive in radiation from the Sun, and thus demonstrated that these planets must have internal sources of heat. In 1969 John Eddy, Pierre Lena, and Robert McQueen flew an infrared interferometer at 12.8 km above the ground on a CV-990 research aircraft from NASA's Ames Research Center to measure the temperature minimum in the solar atmosphere. They used the far infrared range from 80 to 400 μm, as previous work and modelling had shown that the levels of the chromosphere in the temperature minimum region, above the hotter photosphere and below the much hotter and more tenuous corona, should emit in two wavelength ranges, the far infrared and the ultraviolet. They found a value of 4370 K, less than the 5500 K of the photosphere and much less than the million degree regime of the corona. On June 30th 1973 infrared astronomers participated in the longest total solar eclipse in history, flying at 17 km above the Sahara Desert in prototype 001 of the supersonic passenger aircraft Concorde. The speed of the aircraft permitted the observers to see a period of totality lasting 74 min. Six experiments were flown, one investigating the physics of the upper terrestrial atmosphere, four investigating different aspects of the solar atmosphere, and one looking at the scattering by dust and electrons in the interplanetary medium. Figure 3.6 shows the eclipsed Sun on that day taken by Serge Koutschmy, and Fig. 3.7 shows Concorde 001 in the Air and Space Museum at Le Bourget, Paris, with the badge of the trans-African eclipse showing on its side.

The author of this book was one of the scientists on board. His experiment used a helium cooled indium antimonide bolometer to observe the far infrared and submillimetre spectrum of the solar chromosphere, using a Michelson interferometer, through a specially prepared quartz window in the roof of the aircraft. The chromosphere was analysed with sub-arcsecond spatial resolution unequalled until the commissioning of the ALMA array in the high Andes nearly 40 years later.

The main impulse to airborne infrared astronomy was the initiation of NASA's G.P. Kuiper airborne observatory (KAO) in 1975. The observatory was a C141 transport plane, in which was mounted a 91.5 cm diameter telescope, using detectors covering the range from 1 to 500 μm. It flew at altitudes up to 14 km, and therefore above over 99% of atmospheric water vapour. Astronomers who flew in the KAO had to pass a simplified pilot's training course to cover safety aspects, above all possible oxygen escape from the aircraft as the telescope was open to the low pressure outside atmosphere.

Figure 3.8 shows the Kuiper observatory in flight, with its telescope shutter open. The beam from the telescope passed through a quartz window into the observing instrument. The pointing control system had to maintain the telescope locked onto its target with an error of a couple of tenths of an arcsecond while the aircraft orientation changed, rapidly due to turbulence, or more slowly as it flew along its prepared track. Observing flights lasted typically

Fig. 3.6 Image of the Sun during the period of totality of the 1973 solar eclipse. Photo taken bu Serge Kouchmy of the Institut d'Astrophysique, Paris,. During the eclipse, whose totality as observed from the Concorde lasted 74 min, the corona and chromosphere were studied in the mid and far infrared with unprecedented angular resolution. Credit: Serge Koutchmy

some six to seven hours. The main advantage of this observing mode is that different instruments can be taken by research teams and installed on the telescope for a quite limited number of flights. An instrument can use the very latest technology. This can give a big advantage over satellites, which usually take years from concept to flight, and their instruments are normally fixed years before launch. There were many important observations by astronomers using the KAO. Among the most exciting was the discovery of the rings of Uranus in 1977 by James Elliott, Edward Dunham and Jessica Mink (Fig. 3.9).

This was a classical "serendipitous" observation; the observers were not looking for the rings, but were using the occultation of a star to explore Uranus' atmosphere. The rings caused a regular set of dips in the star's light as it passed behind them, and this clue was not hard to interpret. The astronomers knew what they had found by the time the observatory landed. Among

Fig. 3.7 The Concorde 001 supersonic aircraft, which flew to make infrared observations of the 1973 total eclipse, now in the Air and Space Museum, Le Bourget. The eclipse logo can be seen on the side of the Concorde. Credit: Musée de l'Air et de l'Espace du Bourget

Fig. 3.8 Left: The Kuiper Airborne Observatory flying with its telescope door open in 1980. The converted C-141 aircraft had a 36-in. telescope just in front of the wing. Right: Inside the KAO, where the mission crew sat during flight. These consoles were positioned along the side of the aircraft's cabin. The portion of the telescope system that was inside the cabin can be seen at the back of the image. The open telescope cavity was separate from the pressurised cabin. Flying at 14 km above the ground allowed observations across a wide range of the infrared spectrum. Credit: NASA

the other discoveries was the first clear detection of Pluto's atmosphere in 1988, and the observation of disks of gas and dust around stars which emit in the infrared and give important clues about the processes which form planets. Astronomers aboard the KAO first studied the stars around the centre of the galaxy whose orbits have shown in 2018 that they are orbiting a supermassive

Fig. 3.9 Image of Uranus with the IR camera on the Hubble Space Telescope. The rings were discovered in the infrared by observers using the Kuiper Airborne Observatory in 1977. Credits: NASA/JPL/STScI

black hole. In 1987 KAO observations of the supernova SN 1987a in the Large Magellanic Cloud used the infrared range to measure the production rates of the heavy elements such as iron, nickel and cobalt. In 1995 astronomers on the KAO took emission spectra to explore the composition of Mercury's rocks. The KAO left service in 1995 after making over 1400 astronomical flights. Its place was taken by SOFIA, the Stratospheric Observatory for Infrared Astronomy, built by a collaboration between NASA and the German Aerospace Centre, DLR. It was a complex project technically, and did not make astronomical flights until 2010 after some two decades of planning and technical development. SOFIA has a 2.5 m telescope which is mounted in a converted 747SP airliner (Fig. 3.10).

Fig. 3.10 SOFIA the 2.5 m Stratospheric Observatory for Infrared Astronomy a joint project between NASA and the German Space Agency DLR, flying out of NASA's Ames Research Center in California. Left: SOFIA soars over the snow-covered Sierra Nevada mountains with its telescope door open during a test flight. Right: Inside SOFIA during an observing flight at 40,000 ft. The mission crew, including telescope operators and scientists, sit facing the telescope at the back of the aircraft. The portion of the telescope that is inside the cabin is the blue round structure. The beige wall around the blue telescope structure is a pressure bulkhead that separates the open telescope cavity from the pressurised cabin, so the cabin environment feels similar to a commercial aircraft. Credits: left: NASA/Jim Ross, right: NASA/DLR/Fabian Walter

It needs precision guidance for the telescope to overcome the severe demands of air turbulence while remaining locked onto the astronomical object in flight. Both the KAO and SOFIA have been described as "flying organ pipes". Before flight the air in the observers' cabin is cooled so that the telescope does not distort when it reaches the low temperatures at its observing altitude of some 13.7 km above sea level. Before landing the observing door is closed and the telescope enclosure is flooded with dry nitrogen, to prevent condensation on the telescope mirrors. Even so the mirrors need re-aluminizing twice per year. Among the observations carried out so far with Sofia are an analysis of the composition of Pluto's atmosphere and a comprehensive study of the stars and warm dust around a supernova remnant close to the Galactic centre, using the mid infrared radiation emitted by the warm dust which can pass through the clouds of cool dust between us and the centre of the Milky Way. A combined view of these observations, together with contributions from the Herschel Space Observatory (see below), and the Spitzer Space Telescope (see below) are shown in Fig. 3.11.

SOFIA is a vigorous ongoing programme with too many separate results to present in this summary. There are over 700 articles in the professional literature based on SOFIA observations, on subjects as varied as planets in the solar system, protoplanets around other stars, molecules in molecular clouds,

Fig. 3.11 Composite infrared image of the centre of the Milky Way, spanning over 600 light-years across. It helps us to learn how many massive stars are forming in that zone. New data from SOFIA taken at 25 and 37 μm, shown in blue and green, is combined with data from the Herschel Space Observatory, shown in red (70 μm), and the Spitzer Space Telescope, shown in white (8 μm). SOFIA's view reveals features that have never been seen before. Credits: NASA/SOFIA/JPL-Caltech/ESA/Herschel

supernova remnant analysis, active galactic nuclei, regions of massive star formation in the Galaxy, and interstellar magnetic fields. SOFIA observations recently showed that there is considerably more water on the Moon than previously thought, a potentially important finding for the future of space colonisation.

3.5.2   Infrared Astronomy Satellites

3.5.2.1   IRAS: *The Pioneer*

Infrared astronomy from space was initiated by the launch in 1983 of the IRAS satellite, a joint project between NASA, the UK's SERC, and NIVR of the Netherlands. Figure 3.12 depicts IRAS in orbit; it was characterised by the gold-plated sunshield, designed to limit heat input. IRAS had only a small, 57 cm diameter, telescope, cooled to 2 K using liquid helium in order to reduce its own infrared emission to a minimum. It observed in four wavelength bands: at 12, 25, 60, and 100 μm which as we saw previously are not observable from the ground, and cover a wide range of warm to cool objects in space. It had two instruments at the focal plane of the telescope: a low resolution spectrometer, and a photometric imager. The small diameter, limited by launch cost capability, was not a real problem in collecting enough infrared flux from the objects observed, (over 350,000 new sources were discovered) but limited the angular resolution to some 30 arcseconds at 12 μm and to 2.5 arcminutes at 100 μm. The need to cool with liquid helium was a severe limitation on the useful life of IRAS, as the small volume carried had evaporated after 10 months, and observations ceased, although the satellite is still in orbit.

Fig. 3.12 The IRAS infrared astronomical satellite (artist's impression), featuring the protruding sunshield. Credits: NASA/SERC/NIVR

The most powerful contribution of IRAS was its large general source catalogue which included nearly 100,000 infrared-bright galaxies. Follow-up of all the sources over a wide wavelength range has continued to the present day, and some 20% are still unidentified. Many of the infrared galaxies have been studied in detail and the brightest are referred to as LIRGS (Luminous Infrared Galaxies) or even ULIRGS (Ultraluminous Infrared Galaxies). These objects are typically the sites of abundant star formation, shrouded in dense dust clouds, which convert the intense optical and UV emitted by the massive young stars into infrared radiation. Spectroscopy of these galaxies gives us their redshift and hence their distances from us. As they are so intrinsically bright we can detect them at relatively large distances, and their study has given us a way to derive the rate of star formation in the universe as a function

Fig. 3.13 Galactic cirrus. A thin veil of interstellar dust, within the clouds of hydrogen is observed here in the far infrared as photographed by the IRAS satellite. The dust (and the gas) is concentrated towards the plane of the Milky Way, with specific whorls where stellar activity in the plane is stirring it and pushing it outwards. This very cool dust is almost completely invisible at optical and near infrared wavelengths, but emits in the very long wavelength infrared. Credits: NASA/SERC/NIVR

of distance, which is effectively a function of time. We know from this type of infrared studies that the maximum star formation rate occurred when the universe was about one third of its present age. Among the other interesting discoveries by IRAS was the presence of widespread far infrared emission from dust within our Galaxy, extending over the whole sky. Figure 3.13 shows the galactic plane in the 100 μm waveband, with the cirrus: filaments and blobs of infrared emission extending to quite high galactic latitudes.

This cirrus presents problems for the observation of faint extended sources, especially the cosmic microwave background. A striking example of how the infrared contributes to our perception of the interstellar medium is shown in Fig. 3.14 where we can compare a photograph in visible light of the sky containing the constellation of Orion with the corresponding "photograph" made by converting the infrared brightness at 100 μm into a scaled colour image, with yellow corresponding to higher brightness and red to lower brightness. The strongest emission comes from the Orion Nebula, in the lower centre of the image.

IRAS observations were made for stars and galaxies by fixed pointings to source positions on the sky. By later studying the images to find moving sources Jack Meadows and his group in the UK discovered three asteroids, six comets, and a long trail of dust associated with comet Tempel.

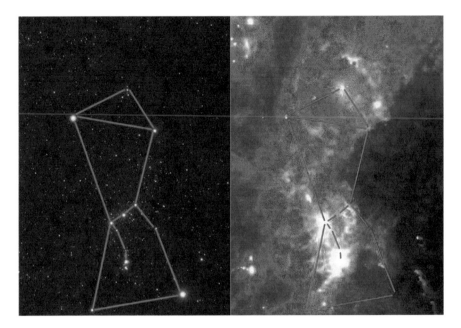

Fig. 3.14 Comparison of the sky around the constellation of Orion in the visible (left hand panel) and in the far infrared from IRAS (right hand panel. The far infrared shows emission from interstellar dust at temperatures below 100 K except around hot stars, where it can be at several hundred K. Credits: NASA/SERC/NIVR

3.5.2.2 ISO

The successor to IRAS was ISO, a joint project between ESA, NASA, and the Japanese space organisation now known as JAXA, launched in 1995. It contained a 60 cm diameter telescope cooled to 1.7 K by liquid helium, and four instruments: ISOCAM, a high resolution camera imaging from 2.5 to 17 µm wavelength, ISOPHOT, a photometer to make calibrated measurements of object luminosities between 2.4 and 240 µm, SWS a spectrometer covering the range 2.4–45 µm, and SWS a spectrometer covering the range from 45 to 186.8 µm. The liquid helium on ISO lasted for just over 2 years, and the mission ended in April 1998. In a little over 900 24/hour orbits it made over 26,000 observations. We can survey the kinds of observations it made during this time. Water vapour was detected in many star-forming regions and in particular in the most studied region: the Orion Nebula, in regions around stars close to their final phases, in the atmospheres of planets in our Solar System, and close to the centre of the Milky Way. Its spectrographs gave major advances in our knowledge of the chemistry of the atmospheres of a number of Solar System Planets. With ISO astronomers were able to study the earliest stages of star formation detecting pre-stellar cores, the densest zones of molecular clouds, which give rise to protostars and then stars. ISO made a detailed study of the ULIRG Arp 220, which is the result of the merger of two galaxies, and whose huge infrared output was first discovered by IRAS.

3.5.2.3   Spitzer

The next step after ISO in the NASA space programme, an 85 cm helium cooled telescope, was originally called SIRTF, but on launch in 2003 it was renamed Spitzer, after the famous US astrophysicist who had given a major impulse to astronomy from space. The optics were of beryllium, which is lightweight and strong, and whose temperature was held below 6 Kelvins (-267 °C). It could take photometric images between 3 and 180 μm wavelength, it could take spectra between 5 and 40 μm, and broad band spectrophotometry between 50 and 100 μm. The 50 kg of liquid helium launched to keep Spitzer's mirror cold had all evaporated by 2009 so from May 15th of that year NASA decided to maintain Spitzer in operation but in a new phase "Spitzer Warm" with the main telescope mirror operating at a temperature of 28 K (–241 °C) maintained at that temperature by radiating its heat to the cold of outer space. It has been operating in that mode ever since, and its cameras operating in the wavelength ranges at 3.6 and 4.5 μm are just as sensitive as before, although the sensitivity at longer wavelengths is clearly reduced. Spitzer has left astronomers with a number of "Legacy" programmes which have formed the basis of research over a much wider range of the spectrum than merely the infrared. These are SWIRE: a survey of infrared objects over a wide area on the sky, GOODS: a very deep survey image of a field originally pioneered by the Hubble Space Telescope, HDFN (the Hubble Deep Field North) and another southern deep field, CDFS (Cosmos Deep Field South), SINGS: detailed infrared maps of nearby galaxies, COSMOS: a combined survey by the HST and Spitzer, and SAGE: studying star formation in the Large Magellanic Cloud.

Spitzer is an observatory satellite, which started with a number of programmes for the scientists who had given years of their scientific lives to the project, and was then opened up to the whole astronomical community. It has made contributions over the full range of astronomical fields: stars, exoplanets and exoplanet formation, galaxies, and cosmology. Here we will pinpoint some key results to give the scope of its work. Figure 3.15 shows a region of star formation in the Milky Way, W3 taken by Spitzer. The youngest stars are at the tips of trunk-like features around the rim of the cavity which has been formed by a blow-out from a previous generation of hot stars.

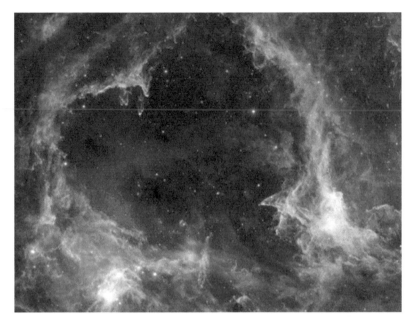

Fig. 3.15 Spitzer image of the star forming region W5 within the Galaxy. The youngest stars are forming at the tips of the columns within the cavity, which was blown by the wind and explosion of one or more massive stars from a previous generation. The blue stars within the cavity are the less massive stars from that generation. Credits: Lori Allen, Xavier Koenig et al. (Harvard-Smithsonian CfA)/JPL-Caltech, NASA

The oldest stars, less massive and therefore longer lasting that those which blew the cavity are the blue dots inside it. Even in the infrared the dust in the star forming clouds around the cavity is not transparent, because the clouds have been compressed by the blast. This is a typical sequence of star formation where one generation breeds another by compressing surrounding clouds of gas.

As an example of the kind of contribution that Spitzer could make we can see in Fig. 3.16 an image of a part of the Pleiades taken with its IRAC camera, combining data at 3.6 and 4.5 μm wavelength, depicted in false colour. The Pleiades is a young star cluster, in which the hot stars have mostly blown away the placental molecular cloud which gave them birth, mostly but not completely. This infrared view shows the hot stars partly shrouded in the dust remaining from the molecular cloud, which does not show up in the visible. A particularly interesting feature of this image is that we can just make out the brown dwarf Teide 1, and part of the image has been reprocessed so that we can see it well. Brown dwarfs are objects with masses less than those of normal stars, but greater than those of planets, a sort of "missing link" between them (but not in an evolutionary sense). Their cores are not hot enough to sustain

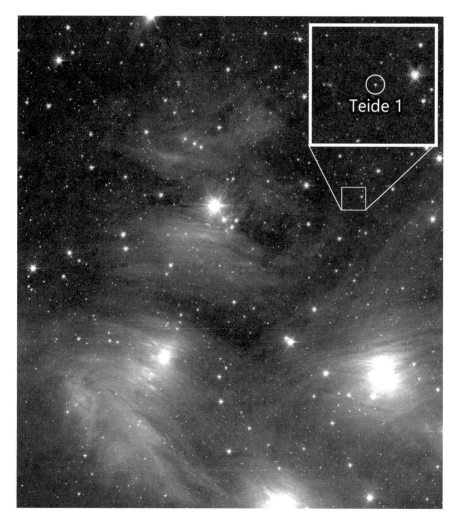

Fig. 3.16 The Pleiades imaged by the Spitzer Space Telescope's IRAC camera; we can see the warm dust from the remains of the star forming cloud as well as the brightest stars. To the top right is the brown dwarf Teide 1, relatively brighter in the infrared than at optical wavelengths. Credits: NASA/Meli thev

nuclear processes, so they shine by the energy they emit due to their own steady contraction under gravity (for stars this contraction phase is what heats the cores, and as the mass is sufficient, the cores reach the temperatures needed for hydrogen to fuse to helium, and the stars stop contracting, remaining in this phase while their hydrogen lasts; they are then on the main sequence of the HR diagram).Even the coolest stars have surface temperatures well over 3500 Kelvins, while for brown dwarfs the temperatures are below 3000 K. Teide

1 has a surface temperature of 2600 K, and a mass one twentieth that of the Sun, but sixty times the mass of Jupiter. Because the luminosities of brown dwarfs are typically less than a thousandth that of the Sun, they are difficult to find, particularly in the optical. An eye check using Fig. 3.1 tells us that in the visible range, just below 1 μm wavelength, the ratio of brightness of two stars with the same size, one with a surface temperature of 5600 K (that of the Sun) and another at 2600 K (you need to interpolate!) will be over one million, but at 4 μm wavelength this ratio is around 100. So the infrared is the best place to look for brown dwarfs. Nevertheless Teide 1, the first brown dwarf to be reliably identified, was found in 1995 with persistent searching by Rafael Rebolo, Maria Rosa Zapatero, and Eduardo Martín using an optical telescope. This was the IAC80, a small instrument at the Teide Observatory of the Instituto de Astrofísica de Canarias, in Tenerife. Since then, as infrared techniques have been used consistently, significant numbers of brown dwarfs have been found in a number of fairly nearby young star clusters in the solar neighbourhood of the Milky Way.

In Fig. 3.17 we show the aftermath of a stellar death: the supernova remnant Cassiopeia A. We will see this object at a variety of wavelengths in several of the chapters of this book, and each wavelength gives us additional information about the physical processes going on within this highly energetic rapidly expanding almost spherical cloud.

The fact that it emits near infrared radiation shows that it contains dust, heated by the radiation first emitted at much shorter wavelengths by the hot gas in the exploding star and its surroundings. Studying in the infrared we learn about how this dust is produced by condensing heavy elements produced within the supernova. We know that when a supernova explodes within a gas cloud it has the effect of dissipating the cloud, thus quenching further star formation in that cloud. However if the same expanding gas reaches the edge of a nearby cloud it can compress it, and trigger new star formation. This is essentially what we were seeing in Fig. 3.15.

A particularly attractive way of looking at global star formation processes over a full galaxy is to see where the dust is being heated by its hot young luminous stars. Figure 3.18 is Spitzer's near infrared image of the galaxy M51, which is a gas-rich star-forming spiral in the local universe. The colours here are not real, but as in Fig. 3.17 are near infrared bands transformed to the visible so that we can get a good idea of the structures. In general terms red means cooler and blue means hotter. The emission from the disc of M51 is mainly from dust, with the arms showing up strongly as pink interwoven strands, with the dust distributed within the gas around the hot star clusters, which show as whiter compact "blobs" of mixed radiation from the stars and

Fig. 3.17 Spitzer combined image of the supernova remnant Cassiopaeia A, in three of the passbands of its MIPS instrument. False colours: 3.6 μm, blue, 4.5 μm green, 8.0 μm red. Credits: NASA/JPL (Whitney Clavin)

surrounding dust. The central zone of the galaxy looks bluer, because there is less gas and dust so we see mainly radiation from the stars. This is also the case for the companion galaxy seen at the top of the image, which has relatively little gas and little star formation so we see the light directly from its stars.

The warm dust is the main feature of many of the Milky Way-like galaxies we see around us when they are imaged by Spitzer in the infrared. A classical example is provided by our neighbour, the Andromeda Galaxy, M31, which is shown in Fig. 3.19. It is particularly interesting to compare this with the "standard" picture of the Andromeda galaxy in the visible, which we showed in Chap. 1, Fig. 1.7, and we will see another version in the ultraviolet, in Chap. 4, Fig. 4.10.

Because Spitzer sees far into the infrared it is also used to obtain information about the galaxies towards the time of their formation. This is because we

Fig. 3.18 The nearby spiral Galaxy M51, the Whirlpool Galaxy, and its interacting companion NGC 5194 imaged in the infrared by the Spitzer space observatory. What we see is not the stellar population but the warm dust in the interstellar medium reflecting the structure of the gas in the medium. Credits: NASA/JPL

Fig. 3.19 Our neighbouring galaxy in the local group, the Andromeda Galaxy, M51, imaged by Spitzer at 24 μm wavelength. The spiral structure is very nicely shown because the warm dust detected at this wavelength is concentrated in the spiral arms where most star formation is occurring. We can also see clearly that the disc is warped, and distorted at the right possibly due to interactions with its satellites. Credits: NASA/JPL

see the universe at $1/(1 + x)$ of its present age by observing light at $(1 + x)$ of its emitted wavelength, as caused by the cosmic expansion. The simplest example is that when $x = 1$ we see galaxies as they were when the age of the universe was ½ of its present age, and the wavelength of its light is twice as big as when it was emitted. This shifting of light to longer wavelengths, known as the "redshift" means that visible light is shifted into the infrared. An example of this is shown in Fig. 3.20 where we see a nearby galaxy and a distant galaxy in the same field on the sky. The conversion of infrared to optical light is the same across the whole image.

We can see that the distant galaxy is a very red blob. It is a blob because Spitzer's mirror was not very big, so the resolution on the sky was limited. It is very red because of the cosmological redshift effect just described. There is also a cosmological dimming effect. If an object has a redshift of x (so that the wavelength of any observed line in its spectrum is $(1 + x)$ times the value it would have if it were here at rest) its surface brightness will be $1/(1 + x)^4$ of the

Fig. 3.20 Two galaxies imaged in the long wavelength band of Spitzer, with identical conversions of infrared to visible light for both. The distant galaxy looks fainter and deep red because its light is "red-shifted" and dimmed due to the expansion of the universe. The image is "pixelated" due to the large pixel size at this long wavelength. Credits: NASA/JPL

value of the same galaxy nearby. So a galaxy with a redshift, x, of 2 will be $(1 + 2)^4$, which is 3^4, which is 81 times fainter than the same galaxy nearby. Fortunately many galaxies in the redshift range greater than 2, i.e. which we detect when the universe was one half or less of its present age, were producing stars at a much greater rate than today, so we can detect them.

One of Spitzer's goals was to try to trace the population of galaxies back to redshifts greater than 3, and in particular to help identify how the rate of galaxy formation has varied with time. The SWIRE survey by Spitzer confirmed that the peak of the galaxy formation rate took place close to z = 2, when the universe had around one third of its present age, and the peak of the population of quasars is found to be around z = 3, when the universe had one quarter of its age.

Spitzer was able to apply its infrared "eye" to many fields of astrophysics. One of the burgeoning recent fields is that of exoplanets, systems of planets around other stars. Perhaps the most striking example of observations by Spitzer is that of Trappist-1 a local star in the Milky Way. This is a very cool dwarf star, coded for temperature by astronomers as M8, which puts its surface temperature at 2700 K (compared with the Sun's 5500 K). In 2015 Belgian astronomers using quite a small telescope ("TRAPPIST") at the La Silla observatory in Chile had detected three Earth-sized planets around this

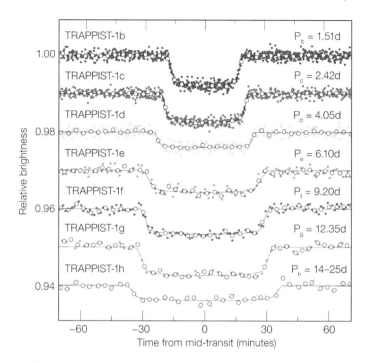

Fig. 3.21 Light curves of the planets around Trappist-1 as observed by the Spitzer Space Telescope. Each curve is produced when one of the planets moves across the face of the star and blocks a small fraction of its infrared light. From these curves estimates of each planet's size and its orbital period can be inferred. Credits: NASA/Monthly Notices of the Royal Astronomical Society, Vol. 475, p. 3577, 2018, Figure 3/L. Delrez et al.

star using the method of transits, in which a small dip in the star's luminosity is detected whenever a planet passes between us and the star. In February 2017 the discovery of four additional planets around Trappist-1 was announced, as a result of work by Spitzer, as well as the ground based VLT (very large telescope) also in Chile. Spitzer had an important advantage in this work, its infrared sensitivity was optimal for the light from the cool star, so it was particularly suited to detecting the four extra blips from the newly discovered planets.

In Fig. 3.21 we show the light curves of the star observed with Spitzer, where we can see clearly in all cases the reduction in the light received from the star as each planet passes in front of it. From the magnitude of this reduction, as well as from the time taken for the light to diminish, we can estimate the size of a planet, while from the time taken for the planet to pass in front of the star we can make an estimate of its orbital parameters. If it is also possible to measure the velocity changes in the star caused as the planets rotate

Fig. 3.22 Artist's impression of the seven planets discovered around the star Trappist-1 four of them by the Spitzer IR space telescope. Credit: NASA/JPL

around it we can estimate their masses. In Fig. 3.22 we have an artist's impression of the seven planets of Trappist-1, derived using the information outlined. Of particular interest is the fact that some of these planets lie in the habitability zone of the star, the zone between whose inner and outer limits liquid water can exist on the surfaces of the planets.

Detailed work on these Trappist-1 planets by the Hubble and Spitzer Space Telescopes has yielded many of their global properties: not only their orbital periods, which are very short, (between 1 and 18 days, showing that they are all close to the star) but also their sizes and mean densities. All of the sizes and densities are in a range close to those of the Earth. However estimates of the amount of far ultraviolet radiation reaching them from their star discourage ideas about expecting signs of life, and a search for signals by the SETI project has not come up with any positive results.

3.5.2.4   The Herschel Space (Far Infrared) Observatory

The Herschel Space Observatory was designed to observe the far infrared; it was essentially a project of ESA, the European Space Agency, with collaboration from NASA, launched in May 2009. It was the largest infrared telescope put into space, with a primary mirror diameter of 3.5 m, necessary because the longer the wavelength the larger the mirror needed to achieve a given angular resolution on the sky. Figure 3.23 is a picture of the Herschel Space Observatory against a background image of the interstellar medium imaged by it.

Fig. 3.23 Herschel Space Observatory against a background image of the interstellar medium. Credit: European Space Agency (ESA)/A. Le Floc'h

The light received by the mirror was focused onto three instruments, all cooled with liquid helium to below −271 K (some 2 degrees above absolute zero). This was an absolute requirement for the long wavelength range, and when the 2300 l of liquid helium had boiled away in 2013 the useful mission lifetime ended. The instruments were PACS: an imaging camera and low resolution spectrometer covering the spectral range from 55 to 210 μm, SPIRE: an imaging camera and low resolution spectrometer covering the range from 194 to 672 μm, and HIFI: a heterodyne detector working on the same principles as a radio receiver and giving extremely high spectral resolution in two bands between 157 and 625 μm. We can see that Herschel operated at much longer wavelengths than the previous infrared satellites, extending its range into what are usually termed submillimetre wavelengths. This is where

Fig. 3.24 Herschel's view of the plane of the Milky Way, a combination of images between 55 μm and 672 μm wavelength, in false colour. Red is the coolest, densest insterstellar dust, the other colours are warmer dust heated by light from star forming regions. Cooler denser clouds of gas and dust are where the next generation of stars will form. Credits: ESA/PACS/SPIRE/S-Molinari, Hi-GAL project

infrared astronomy and radioastronomy meet and overlap. We usually refer to the measurements made with conventional broad-band detectors as far infrared, while those made with heterodyne detectors are referred to as submillimetre. To avoid problems with the radiation from the Earth and its upper atmosphere Herschel was put into a high orbit, at a "Lagrangian point" 1.5 million kilometres from Earth, a point where it remained stable and not too distant from the Earth, but not orbiting around it.

There have been well over 2000 professional publications, and some 140 PhD theses produced on the basis of Herschel data; here we will see only a few highlights to give a feel for what has been accomplished. In Fig. 3.24 we see Herschel's view of the plane of the Milky Way. It is a combination of data from PACS and SPIRE, covering a span of 12° in the sky along the Galactic Plane, and just over 1° above and 1° below the plane. These instruments show in the longer wavelengths (depicted as reds) complex filaments and clouds of cool dust, which reveals the structure of the denser parts of the interstellar medium, and in the shorter wavelengths (depicted in green, blue and white) the warmer dust surrounding associations of young stars.

Detailed studies of these regions, comparing Herschel data with Spitzer data, with the distribution of neutral atomic and molecular gas from radio measurements, and with optical imaging, enable us to study sequences of the behaviour of the interstellar medium before, around, and after the star formation process. The time for the condensation of interstellar gas into stars is very short in astronomical terms, so astronomers are eager to find material in all of these processes to construct valid theories of star formation, which are still very incomplete. In Fig. 3.25 we see a zone of massive star formation, Cygnus X, in the direction of the constellation of Cygnus on the sky. This is also a

Fig. 3.25 Herschel far infrared image of Cygnus-X, a strong star forming region. White, blue and yellow coded regions show the hot and warm dust around clusters of young stars. The red zones are cooler denser clouds of gas and dust where the next generation of stars will form. Credit: ESA/PACS/SPIRE/Martin Hennemann & Frédérique Motte, Laboratoire AIM Paris-Saclay, CEA/Irfu - CNRS/INSU - Univ. Paris Diderot, France

combined PACS+SPIRE image, and shows two regions where massive hot stars are forming, to the left and the right of the picture, heating the dust around them, and a cooler zone between the two. The deep red clumps distributed in the image are regions of denser gas where the next generation of stars will form.

One of Herschel's prime observations was that of water in the universe. This was a key objective of the HIFI instrument, designed to detect spectral lines at high resolution, including over 40 lines emitted by water molecules in interstellar space and in the solar system. One of the first advances was the detection of water vapour in Lynds 1544, a pre-stellar "core" of dense molecular gas. In this single object there is over 2000 times the quantity of water in the Earth's oceans! The discovery of water in the disc of gas and dust around the star TW Hydrae showed that water can be present in these discs, which are now known to be the places where planets form. Making observations for water throughout our Solar System, Herschel detected water vapour in comets, in the asteroid Ceres, the largest body in the asteroid belt between Mars and Jupiter, and found a huge torus of water around Saturn, associated with its small satellite Enceladus. Later NASA's Cassini space probe detected plumes of water vapour around Enceladus, coming from an underground ocean underneath its icy crust.

Figure 3.26 shows the Herschel image of water vapour around Jupiter, superposed on an image of Jupiter taken by the Hubble Space Telescope, for comparison. There is considerable asymmetry, with more water vapour above the southern hemisphere than the northern hemisphere. The experts believe that almost all of this water vapour was in fact delivered to the Jupiter atmosphere by

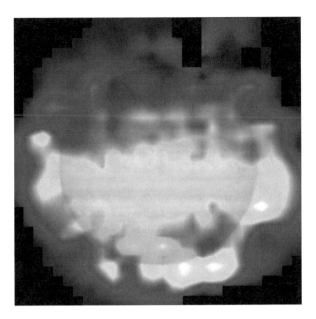

Fig. 3.26 The distribution of water vapour in Jupiter's outer atmosphere. The white zones are those with maximum water concentration, the cyan zones have rather less water, and the blue zones have least water. Notice that there is more water vapour over the southern hemisphere than over the northern hemisphere (see the text for the probable explanation). The map was made with Herschel's PACS instrument, at a wavelength of 66.4 μm Underlying is an image in visible light from the Hubble Space Telescope. Credits: ESA/T. Cavalié et al. (University of Bordeaux/NASA/Reta Beebe) (New Mexico State University/Science Photo Library)

comet Shoemaker-Levy 9 which impacted the southern hemisphere in 1994. One of the current theories about the water found on Earth is that it was brought in by impacting comets, during the formation of the planets, but also in more recent epochs. An interesting check on this has been carried out by comparing the proportion of hydrogen to deuterium found in water on the Earth with the same ratio found in comets and other solar system bodies. These measurements have been carried out with Herschel and a summary of the results is shown in the graph of Fig. 3.27. The principal debate is whether comets or asteroids have been the principal source of the Earth's water. Until the measurements made by ESA's Rosetta probe on comet Churyumov-Gerasimenko in 2015 it was generally assumed that comets are the principal source. In general the measurements in Fig. 3.23 for the comets give D/H values somewhat higher than terrestrial, but the Rosetta measurements gave a value of D/H three times higher than the terrestrial value. So now the view is that the water on Earth was brought here by asteroids.

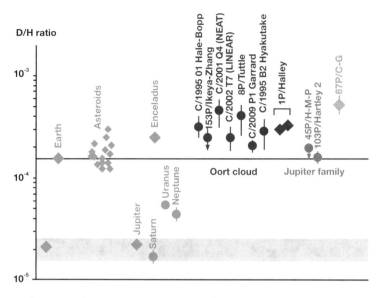

Fig. 3.27 The ratio of Deuterium to Hydrogen found in water on Earth (blue diamond and solid horizontal line) compared with the ratios measured, principally with the Herschel Space Observatory, for planets and moons (blue), chondritic meteorites from the Asteroid Belt (grey), comets originating from the Oort cloud (purple) and Jupiter family comets (pink). Rosetta's Jupiter-family comet is picked out in green. Diamonds show data obtained in situ, and circles data obtained by telescopic observations. The results suggests strongly that the Earth's water was brought here by asteroids during the planetary formation phase. Credits: First measurements of comet 67P/C-G's water ratio. Spacecraft: ESA/ATG media lab; Comet: ESA/Rosetta/NavCam; Data: Altwegg et al. 2015, Science, Vol. 247ᵃ, p 387A (and references therein)

3.5.2.5   The James Webb Space Telescope

Although the James Webb Space Telescope (JWST) will not be launched till late 2021, it will be the most important infrared telescope ever built, and will take astronomical observations to the edge of the universe of stars and galaxies, so we should take a look at its structure and capabilities. The JWST will have a 6.5 m gold surfaced primary mirror, made up of 18 hexagonal segments, as shown in Fig. 3.28. It will be in some sense the successor of the Hubble Space Telescope, but instead of observing ultraviolet plus visible range it will observe from the visible, at 0.6 μm to the mid infrared at 27 μm. The infrared range demands the larger mirror to achieve good angular resolution, and it must be cool to minimise the infrared emitted by the telescope itself. This will be achieved by placing it at a Lagrangian point with a large sunshield shown in the figure. To launch a telescope of these dimensions within the envelope of

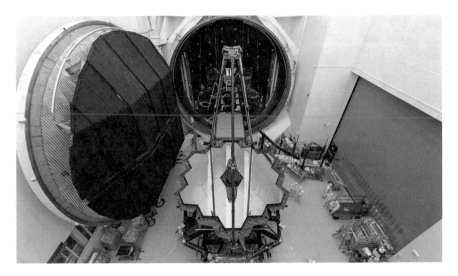

Fig. 3.28 The James Webb Space Telescope, in an early laboratory test phase. It is due for launch in October 2021. Credits: NASA/Northrop Grumman Ball Aerospace

any launcher it had to be designed as a folded system which unfolds in space. The sunshield, which will have a total span of 22 m but which cannot be too massive, is made of many sheets of a polyamide film coated against reflection with aluminium on one side and silicon on the other. This has given very tricky engineering problems on test, which have been the main cause of the delays in the programme which have put back the launch date by several years. The planned duration of the mission is 5 years, with possible extension of up to 10 years.

The JWST will have three main instruments: NIRCam, an infrared imaging camera covering the wavelength range from 0.6 to 5 μm, NIRSpec, a spectrograph covering the same range, and MIRI, a mid-infrared instrument, with sensitivity from 5 to 27 μm, capable of both imaging and spectroscopy. There will also be a fine guidance sensor FGS to guide and stabilise the telescope during observations, into which a slitless spectrograph NIRISS is incorporated.

The mission of the JWST has four main goals: to find and characterise the earliest stars and galaxies formed after the Big Bang, to study the formation and evolution of galaxies in general, to study the formation of stars and of planetary systems around them, and to try to investigate aspects of the origins of life. Although as I write this the launch date is still some time away, 13 programmes, with 460 hours of early observation time, were adopted in 2017, covering the solar system, exoplanets, stars and star formation, local and distant galaxies, gravitational lenses, and quasars.

Further Reading

Rieke, George: History of infrared telescopes and astronomy. Experimental Astronomy, Volume 25, article, 125, (2009). Springer Link

Walker Helen: A brief history of Infrared Astronomy. Astronomy & Geophysics, Volume 41, Issue 5, p. 10

Beckman, John E. and Moorwood, A.F.M. Infrared Astronomy. Reports on Progress in Physics, Vol. 42, No 1. P. 87 1979

Clements, David F. : Infrared Astronomy: Seeing the Heat, CRC Press, 2014

Rottner, Renee "Making the Invisible Visible" A History of the Spitzer Infrared Telescope Facility, NASA. Office of Communications NASA History Division Washington, DC 20546

4

Ultraviolet Astronomy

4.1   Why Study Astronomy in the Ultraviolet?

The most important reason to observe astronomical sources in the ultraviolet range of the spectrum is that some of the key spectral features of the most abundant element, hydrogen, are found there. These include the shortest wavelength series of lines produced by atomic hydrogen, discovered in the laboratory by Harvard physicist Theodore Lyman between 1906 and 1914, and the associated continuous spectrum at even shorter wavelengths. As hydrogen is by far the most abundant element in the universe studying these lines and the continuum in both emission and absorption are essential tools for exploring the behaviour of stars, galaxies, and the interstellar medium.

In Fig. 4.1 we show this series of lines, called the Lyman series, with the longest wavelength line of the series, Lyman-α, at 1215.67 Angstroms (121.57 nm) and the other lines crowding towards the limit at 912 A (91.2 nm) beyond which we find the Lyman continuum. The lines represent the jumps in energy as electrons transit between the energy levels within the atom, and the continuum is the energy emitted (absorbed) when a free electron is captured by a proton to form a neutral atom (or a bound electron escapes from a neutral hydrogen atom leaving it ionised, i.e. a proton and a free electron. These transitions and their associated photons allow astronomers to follow the phenomena of interaction between stars and the interstellar medium which are basic to star formation and indeed to many processes within galaxies. The fact that they are in the ultraviolet part of the spectrum has obliged astronomers to make observations from space, because the atmosphere is opaque to this wavelength range. So virtually all ultraviolet (UV)

© Springer Nature Switzerland AG 2021
J. E. Beckman, *Multimessenger Astronomy*, Astronomers' Universe,
https://doi.org/10.1007/978-3-030-68372-6_4

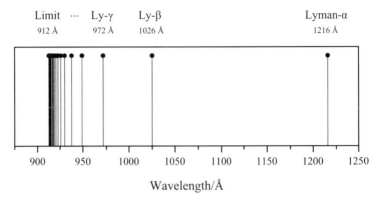

Fig. 4.1 Graph showing the wavelengths of the discrete lines in the Lyman series emitted by atomic hydrogen. Wavelengths are in Angstrom (10 Angstrom = 1 nanometre, nm) These lines are all in the far UV. Beyond the short wavelength limit at 912A continuous radiation is emitted. Credit: Adriferr at en.wikipedia derivative work: OrangeDog (talk • contribs), CC BY-SA 3.0 (https://creativecommons.org/licenses/by-sa/3.0), via Wikimedia Commons

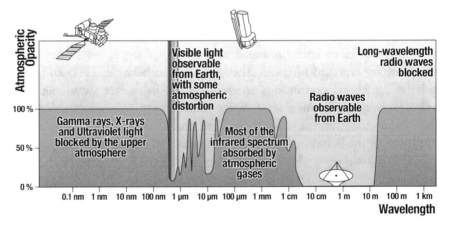

Fig. 4.2 The wavelength regions for which the Earth's atmosphere is opaque or transparent to electromagnetic radiation. The principal regions of transparency are in the visible and the radio wavelength, The infrared has regions of partial transparency, and the atmosphere is opaque to the UV and all shorter wavelengths. In the infrared and all wavelengths shorter than the UV astronomy is best performed from space The graph represents conditions as sea level. Credit: NASA (original); SVG by Mysid., Public domain, via Wikimedia Commons IAC-UC3

astronomy has been feasible only since rockets and satellites enabled telescopes and detectors to surmount the atmosphere.

Figure 4.2 shows the opacity of the atmosphere over a wide range of the electromagnetic spectrum, and we see that at wavelengths shorter than 300 nm all the radiation is absorbed. There is a broad convention to describe the UV range between 400 and 300 nm as the "near" UV (NUV), between 300 and 200 nm as the "mid" UV, and wavelength below 200 nm, ranging down to the X-ray region, as "far" UV (FUV). It is possible to make some observations in the near UV from mountain top observatories, but the mid and far UV must be observed from outside the atmosphere, so here observations from space are essential.

4.2 Ultraviolet from Rockets and Satellites

4.2.1 Early UV Observations from Rockets and Satellites

Here I will give an extremely brief look at the development of UV astronomy since its inception with the first rocket flight to take a UV spectrum of the Sun in 1946. This is anything but exhaustive, and will give only a guideline approach. From 1962 to 1975 NASA launched 8 orbiting solar observatories (OSO1–8) which took thousands of UV spectra of the Sun, and from 1968 to 1981 launched a series of orbiting astronomical observatories (OAO's) to study the stars and the interstellar medium in the range 120–400 nm. The last of these was renamed Copernicus, while on orbit, to celebrate the 500th anniversary of the birth of Nicolas Copernicus, who first placed the Sun at the centre of the Solar System. As well as a UV telescope of 80 cm in diameter the satellite carried an X-ray detector.

4.2.2 IUE

The most successful UV satellite before the Hubble Space Telescope was the International Ultraviolet Explorer, IUE, launched in 1978 by NASA, as a joint project with the European Space Agency (ESA) and the United Kingdom, from where the original idea had come. Its special claim to originality was that it was indeed used as an observatory. European astronomers would travel to the ESA ground station in Madrid, and US astronomers to the ground station at the Goddard Space Flight Center in Maryland, and would observe their own objects in real time, just as they were accustomed to doing at ground-based observatories. The author remembers arriving in Madrid on an overcast day, and for a moment thinking that the observations would be impossible, only to realise immediately that we were in an age of space observatories. The original mission plan was for 3 years, but in fact it lasted 18 years and was terminated only for financial reasons, when the system was still working at

Fig. 4.3 The IUE satellite during a test phase. The telescope tube points out of the body of the satellite which contains the electronics. The slanted aperture at the top of the telescope is the sun-shield. During observations this is held on the sunward side of the satellite. The propellant system and engines have been added at this time. Credits: NASA/ESA/ SERC (UK)

almost peak efficiency. Figure 4.3 is a picture of IUE during its pre-launch test phase.

Over 104,000 observations were made with IUE, on objects ranging from small bodies in the Solar System to distant quasars. IUE was operational during the passage of Halley's comet in 1986 and was used to determine the amount of water vapour lost by the comet, some 300 million tons, as it passed close to the Sun. IUE had two spectrographs, covering respectively the spectral ranges from 115 to 198 nm and 180 to 335 nm. It had quite a small telescope, only 45 cm in diameter, with a moderate resolution on the sky of 2 arcseconds, but its spectral resolution was very good, some 15,000, which means that it had a resolution in velocity of just under 20 km/s. This is considered in the intermediate range for even good spectrographs on ground based telescopes, and it allowed important advances in the physics of stars, galaxies, and the interstellar medium.

4.2.3   The Far Ultraviolet Spectroscopic Explorer (FUSE)

The Far Ultraviolet Spectroscopic Explorer (FUSE) was not a general purpose observatory, but was designed specifically to observe in the wavelength range from 90.5 to 119.5 nm, with one of its main aims to observe deuterium in the nearby parts of our Galaxy, which cannot be detected at optical wavelengths. FUSE was operational for 8 years, from 1999 to 2007, and the data gave rise to over 400 scientific papers, on subjects ranging from deuterium to the properties of the atmospheres and winds of hot stars, and diagnostics of exoplanets. A picture of FUSE in the cleanroom at NASA's Godard Space Flight Center is shown in Fig. 4.4.

4.2.4   The Galaxy Evolution Explorer (GALEX)

GALEX, the Galaxy Evolution Explorer was, in common with all of NASA's astronomy missions termed Explorer, a small relatively cheap mission. It flew a 50 cm telescope whose camera had two photometric imaging channels, the NUV from 278 to 177 nm, and the far UV from 178 to 134 nm. Its special feature was its large imaging area on the sky, a very large 1.2° × 1.2°, two orders of magnitude bigger than the Wide Field Camera on the Hubble Space Telescope, of 2.3 arcmin × 2.1 arcmin. This comparison shows that GALEX was intended as a survey instrument, to cover the full sky, while the HST is far more specialised on individual objects and small fields, measured with high angular resolution. GALEX flew from 2003 to 2013 and one of its main mission goals was to measure calibration fluxes from nearly 1000 local stars to be used for future UV observations of all types of objects. Another prime goal was to image whole galaxies in the UV, to measure the contributions of their hot stars.

4.2.5   The Hubble Space Telescope (HST) in the UV

Although the HST has achieved maximum fame because of its wonderful optical resolution, yielding a constant cascade of memorable images, we should not forget that one of its prime scientific aims was to extend both imaging and spectroscopy of astronomical objects into the UV. There have been three generations of UV spectrographs on HST. The first was the High Resolution Spectrograph (HRS), which could observe much fainter objects than IUE, but whose spectral resolution was not so good, due to the rather

Fig. 4.4 The FUSE satellite in preparation in a clean room at NASA's Kennedy Space Center, Florida. Credit: NASA image and video library

poor figure of the telescope's primary mirror. In 1999 the HRS was replaced by the Space Telescope Imaging Spectrograph (STIS), which was able to perform spectroscopy in both FUV and NUV, and led the way in high resolution spectroscopic observations of the interstellar medium, element abundances is stars, and the spectroscopy of exoplanets during transits across their parent stars. It is particularly useful for measurements of the Lyman-α line of hydrogen in objects with low radial velocities (those with higher radial velocities are often observed in the visible, or even the near infrared due to their redshifts), because its high resolution permits the Lyman-α from an astronomical source to be separated from that emitted by the diffuse but hot outer atmosphere of the Earth. STIS had an electrical failure in 2004, but its electronics were replaced by a "spacewalking" astronaut in 2009, and it has been working since then. A third generation UV instrument, the Cosmic Origins Spectrograph (COS) was installed at the same time, and was designed to have maximum efficiency for faint objects by reducing the number of optical components to a minimum.

4.3 Specific Studies and Results of UV Astronomy

4.3.1 The Sun

The contribution of UV astronomy to solar research has been considerable. The reason is that the tenuous outer atmosphere of the sun, the corona, is hot, much hotter than the visible surface below it, reaching temperatures of over a million Kelvins. At these temperatures all the atoms are multiply ionised, they have lost several electrons. The emission spectra of these ions has most of its lines in the UV or even in the X-ray range. There is a lot of interest in diagnosing the physical state of the corona because through it flow the high energy particles which form the solar wind and which impinge on the outer atmosphere of the Earth. This interaction affects communications on Earth, but the high energy particles also affect the environment of manned space probes, and this "space weather" is of considerable interest to NASA and the other space agencies. The particles also have long term effects on the crews of aircraft. These are subjects of direct medical study, but the underlying physics of the solar wind is best understood by studying the corona from which it flows. NASA has launched a dozen probes over the years to study the sun, and some half of them have taken UV images. Among the earlier were those taken by instruments on board the Space Shuttle. The longest and most productive UV imaging mission has been the Solar Dynamics Observatory, SDO launched in 2010 and still in normal operation. It takes images at discrete wavelength between the mid-UV at 160 nm wavelength and the extreme UV at only 10 nm, close to the X-ray range. Figure 4.5a, b shows two images from the cameras of the SDO.

The phenomena observed and measured include those associated with strong magnetic activity around the sunspots in the photosphere and chromosphere below the corona. Escaping electrons, protons and alpha particles are detected at velocities of hundreds of kilometres per second, impelled by the release of stored magnetic energy. This is highlighted in Fig. 4.5a, which shows an active magnetic zone taken by SDO at 17 nm wavelength. A more global picture of the corona from SDO is shown in Fig. 4.5b where the most striking features are the dark areas, discovered with X-ray probes in the 1960s and 1970s. These are known as coronal holes, and are zones where the corona is less dense, allowing a steady stream of particles to escape into interplanetary space forming a steady solar wind. A given hole may last for months, producing predictable interactions, such as aurorae, in the Earth's magnetosphere. During the minimum phase of a solar activity cycle, the coronal holes provide all the solar wind. But during the more active phases localised coronal outbursts, such as that imaged in Fig. 4.5a cause sharp short term spikes in the wind, which can have strong impact on communications, notably by short wave radio. The NASA-ESA Solar Orbiter Probe has UV cameras which can support high temperatures and which will provide complementary information about the solar atmosphere to the NASA Parker solar probe. This probe, launched in 2019, will take a packet of instruments close to the outer corona and will attain temperatures no known camera technology can survive, so the Solar Orbiter will provide the simultaneous UV imaging to complete the overall mission.

4.3.2   Comets, Planets and Exoplanets

Comets, planets, and their satellites, have been observed in the UV by IUE, FUSE, and HST. FUSE was used to observe carbon monoxide formed from the dissociation of carbon dioxide as the comets approached the Sun. Absorption by water vapour and molecular oxygen were observed by the ALICE UV spectrometer on board the ROSETTA mission to comet Churyumov-Gerasimenko, reported in 2017, and sulphur carbide, the hydroxyl molecule, and molecular sulphur have all been identified in the UV spectra of comets. The first studies of the aurorae on planets other than the

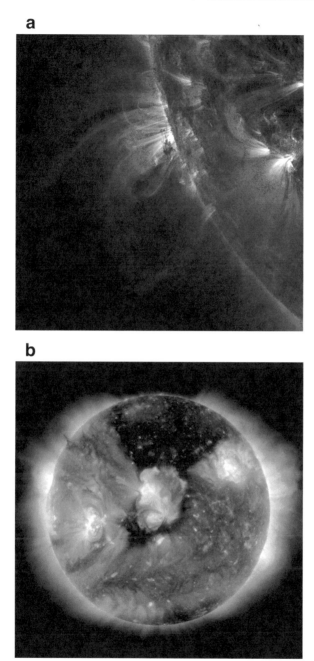

Fig. 4.5 (a) Zone of solar activity captured by NASA's solar dynamics observatory SDO (June 13th 2013) in the far UV at 17.1 nm wavelength The wealth of structures in the sun's upper atmosphere, the corona, is shaped by magnetic fields some of which loop out and back into the solar surface, the photosphere. You can sense the intense motions detected here, and seen in the videos obtained by the satellite. Where the magnetic lines do not loop down the particles are flowing outwards at high speeds to

Earth were made by IUE, as well as by the Voyager interplanetary space probe. Auroral spectra of atomic hydrogen (Lyman-α) and molecular hydrogen were observed with the STIS UV spectrograph on the Hubble Space Telescope near the poles of Jupiter and Saturn, while the FUSE spectra of Jupiter's aurorae contain many lines of molecular hydrogen (Fig. 4.6).

The volcanic activity on Jupiter's nearest large moon Io was observed in sulphur dioxide emission in NUV spectra taken with STIS, and FUSE had observed singly ionised carbon, singly, doubly and triply ionised sulphur and doubly ionised chlorine in the FUV spectrum of Io. Studies of exoplanets are an increasingly popular and practised part of astronomy. The absorption of stellar emission by the atmosphere of an exoplanet as it transits its parent star is used as a powerful probe of its chemical composition and structure. As early as 2003 observers using the STIS UV spectrometer on HST found that they could detect hydrogen in the outer atmosphere of a "hot Jupiter" planet, by measuring the decrease of the star's Lyman-α emission during transit. A hot Jupiter is a massive planet on an orbit close to the parent star. The detection of hydrogen, measurable to quite a long way from the centre of the planet, showed hydrogen evaporating away from its atmosphere at a considerable rate, 10,000 tonnes/s. This type of measurements has been repeated over the years with other similar planets, and other atoms, neutral oxygen and singly ionised carbon, have been measured in their upper atmospheres this way. Similar results have been found for somewhat smaller hot planets, of Neptune or Uranus size, but hot planets of mass similar to Earth do not show this effect, presumably because their smaller mass implies that they have already lost their hydrogen. These examples are illustrative of the kind of methods used to investigate the atmospheres of exoplanets in any of the observable wavelength ranges, but this is a vast subject nowadays, so this is only a tiny indication of what is being done.

4.3.3 Pre-main Sequence Stars and Cool Stars

It may seem odd that stars which have not yet stabilised to steady emission by nuclear burning in their cores, and therefore still "warming up" emit much

Fig. 4.5 (continued) form a part of the solar wind. The local temperatures within the coronal loops can reach 10 million K. (b) The Sun from SDO with a large area coronal hole shown as dark in this far UV image, taken when the Sun was showing an intermediate level of activity. As a substantial part of it is facing the Earth a solar wind component was directed towards us on the day this was taken. Credits: NASA image and video library

a

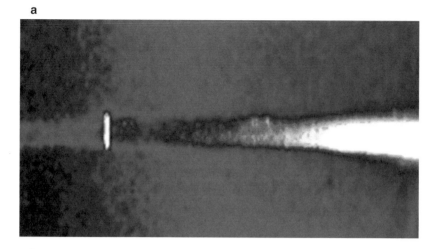

b

Fig. 4.6   (a) UV emission spectrum of the norther aurora on Jupiter, taken with the IUE satellite. The bright bar to the left is the Lyman-a line of atomic hydrogen. (b) The north polar aurora of Jupiter from the Hubble Space Telescope. The aurora, in the far UV is superposed on a previous HST visible image of the full planet. The circular traces in the aurora are magnetic "footprints" caused by Jupiter's satellites Io, Ganymede, and Europa. Credits: NASA/ESA/SERC (UK)

energy in the UV, which as we have said comes mainly from hot material. But in the later stages of formation, cool stars are accreting mass from the surrounding interstellar medium which first falls into a disc of matter rotating around the star, and finally flows onto the stellar surface along lines of magnetic field. The gravitational energy which it gains as it falls is transformed into other types of energy in a magnetic shock region near the surface, among them energy in the UV in both spectral continuum and lines. Among the most widely observed and researched of these stars are the T Tauri stars, named for the archetype of this class, T of the constellation Taurus. An atlas of UV spectra for 50 T Tauri stars and over 70 of the related Herbig As/Be stars from IUE measurements between 115 and 198 nm detected emission lines from triply ionised silicon and carbon, and from fluorescent molecular hydrogen. The improved resolution of the STIS and COS instruments on the Hubble Space Telescope added lines of quadruply ionised nitrogen (nitrogen which has lost 4 of its electrons) and doubly ionised helium. These types of measurements allowed astronomers to quantify the rates of mass accretion during these late stages of star formation, especially when time sequence observation of the same star was possible. The process of star formation has been one of the key puzzles to solve in astronomy and this UV information gives us important insights into it for the low mass range of stars.

We have seen that the UV allows us to map the solar corona and to make estimates of its physical properties, temperature and density, its structure, and its kinematics. The coronae of other relatively small stars like the sun can be observed via their UV emission, and while two dimensional mapping of them is not, for the moment, feasible, their temperatures and densities can be estimated. Coronal temperatures range from a couple of hundred thousand K to over a million K. Multiply ionised iron has many different emission lines in the UV: there are lines at 124.2 and 134.9 nm produced by six times ionised iron, and at 135.4 nm from twelve times ionised iron. These are observable with the UV spectrographs on HST. The former is observed in the coronae of many of the coolest dwarfs, M dwarfs, while the latter only in a few. Twelve times ionised iron is produced by plasma at 10 million degrees, and is expected to be emitted by the coronal loops similar to those we observe on the Sun. Going further into the UV, lines of seventeen and eighteen times ionised iron, at 97.4 nm and 111.8 nm respectively have been detected in the spectra of those cool stars showing coronal variability, with the spectrograph on the FUSE satellite. Cool supergiant stars, as well as cool dwarfs, also have very hot outer atmospheres, and UV spectra allow us to investigate them. HST has allowed observers to detect and measure lines due to neutral sulphur and chlorine, as well as lines emitted by molecular hydrogen and carbon monoxide in these types of stars. All of these lines are formed by an optical "pumping" processes, similar to that which stimulates lasers, with the pumping source of energy being a strong emission line Lyman-β, produced by atomic hydrogen. By measuring the strengths of these lines we can indirectly measure the atomic hydrogen spectrum emitted by the stars. This cannot be observed directly because the neutral hydrogen between us and even the nearest stars effectively absorbs it.

4.3.4   Hot Stars

One of the major fields of study in the UV, to which IUE gave a great impulse, is that of hot stars. The hotter the object, the shorter is the peak wavelength range in which it emits. Hot stars emit a large part of their light in the UV, and IUE played a key role in enhancing our knowledge of these stars. Among these are the white dwarfs, which are the remnants of medium mass stars which have thrown off their outer layers in the form of shells, in nova explosions.

The remaining cores are very hot, with temperatures of up to 100,000 K, but very compact, with radii similar to that of the Earth, even though their masses are similar to the mass of the Sun. So although they are so hot, they emit relatively little total energy, and the fraction of this energy in the visible range is low. They are very faint in the visible, but more detectable in the UV. So it took a UV observatory satellite, IUE, to discover many of them, and in particular to find many white dwarf companions of other known stars. Figure 4.7 shows a white dwarf star at the centre of a "planetary nebula". The nebula was the outer envelope of the red giant star which exploded as a nova, leaving the hot dense white dwarf in the centre. The UV is helping us to understand this late phase of an intermediate mass star. Stars in binaries can be used to measure basic physical properties such as their masses, using the kinematics of their orbits, so these discoveries were important for all branches of stellar astronomy. Another of the important fields of hot star physics is that of the stellar winds of O and B stars. One of the ways in which virtually all stars emit energy, apart from the obvious way by emitting electromagnetic radiation, is by a stellar wind, a flux of particles, principally protons and electrons, but with an admixture of other atoms and ions. The solar wind is now well-known; its physics and it interaction with the Earth's atmosphere have been studied *in situ* by major solar system satellite programmes. But although its local impact is considerable, the rate of mass loss per year in the solar wind is really small, it loses only one part in 10^{14} of its mass per year. The speed of

Fig. 4.7 Planetary nebula the "Ring Nebula" in the constellation of Lyra. The faint dot at its centre is the white dwarf which blew off the shells detected around it, and, whose spectrum was first taken with the IUE satellite. Credits: Daniel López, IAC/ ING

outflow of the solar wind is as high as 750 km/s. But hot stars, with masses ten times that of the Sun, or more, have very powerful winds. Their mass flows outwards at speeds of thousands of km/s and their loss rates can be as high as one part in 10^5 of their mass per year, which if maintained would mean that they would lose all their mass in only 100,000 years, much shorter than even their short lifetimes, which are typically a few million years. So although these rates are not maintained, the lifetimes of the massive stars and whether or not they will eventually explode as supernovae, depend critically on them. Much of the observational work to characterise these winds, and relate them to the evolutionary properties of the massive young hot stars, was pioneered with IUE observations, and continues with data from HST.

4.3.5 UV from Shocked Regions in the Interstellar Medium

Observing the interstellar medium in the ultraviolet can be considered an extension of observing in the visible, in that the processes detected and measured are similar, but at higher energies. In the chapter on optical astronomy we took a brief look at HII regions, the zones of ionised hydrogen around hot, young, star clusters (OB clusters), mentioning that they are used to determine the abundances of the elements by measuring their emission lines in these HII regions. But in zones where the violence of a supernova has had an impact, the UV is of particular importance because the expanding gas from the supernova explosion causes shocks in the interstellar medium which are best observed in the UV or even at higher energies. A beautiful example of this is found quite close to us, astronomically speaking. The Cygnus Loop is the remnant of a supernova at some 2500 light years from us in the direction of the constellation of Cygnus. The explosion took place some 20,000 years ago, as calculated from its present size and rate of expansion, with plausible dynamical assumptions about how the expanding gas interacts with its gaseous surroundings. In Fig. 4.8 we show an ultraviolet image of the Cygnus Loop made by the wide field UV satellite GALEX. We can see that the UV is coming from wispy gas, which is shocked as it impinges on the interstellar medium of its surroundings. Although this continuous shock has slowed down its expansion, velocity measurements on spectral lines show that it is still expanding at some 350 km/s, and feeding energy into the interstellar medium.

4.3.6 The Interstellar Medium and the Cosmic Deuterium Abundance

Among the many contributions of IUE to our wider knowledge of astronomy was the early mapping of the structure of the local interstellar medium, within a few hundred light years of the Sun. There are two strong lines of singly ionised magnesium in the near UV at 279.5 and 280.2 nm which appear as emission lines in the spectra of cool stars similar to the Sun. These lines have complex structure: in the middle of their sharp emission peak there is a trough due to the scattering of photons in the atmospheres of the emitting stars. But with IUE's excellent spectral resolution extra troughs were seen within the emission lines. At first astronomers thought that they were due to complex dynamical structures in these atmospheres, but quite quickly they realised that they were absorption lines caused by the interstellar medium, because

Fig. 4.8 Ultraviolet image of the Cygnus Loop, the expanding remnant of a supernova which occurred some 10,000 years ago, towards the constellation of Cygnus, the image diameter is over three times that of the moon. Taken with NASA's Galaxy Evolution Explorer UV camera. Credits: NASA/JPL-Caltech

they were too narrow to be stellar lines. They are formed in the cooler clouds embedded in the generally hot gas forming the "local bubble" in which the Sun is situated. Using IUE spectra of cool dwarfs and giants in the neighbour-hood of the Sun it was possible to map many of these clouds, inferring their sizes, temperatures and masses. Typically they are at temperatures of around 100K, while the surrounding gas is at temperatures of hundreds of thousands of degrees. It turns out that although these cool clouds occupy only a few percent of the local volume, they contribute over half the mass of the interstellar gas, because they are much denser than the hot gas surrounding them. The magnesium lines played a key role because the lines due to atomic hydrogen are too strong: they satúrate in very short columns of gas and do not allow us to measure how much gas is in a cloud, while the lines of sodium (the well known yellow lines which were common in street lighting) at 589 and 589.6 nm are weaker and do not show up for many of the smaller clouds. The MgII lines are the "goldilocks" lines in this case, strong enough to be detected in all of these clouds, but not strong enough to be saturated. This comprehensive 3D mapping of the local interstellar medium is of great interest, because its density and temperature structure can be used as a template to help understand the structure of the interstellar gas under the very wide variety of conditions found throughout our Galaxy and in other galaxies. It is analogous to our use of the Sun as a template for understanding the stars. We know that there is great variety in the stars and in the interstellar medium, but these local measurements give us a basic understanding on which to build (Fig. 4.9).

One of the most powerful motives for space ultraviolet astronomy was the search for deuterium in different types of astronomical objects. Deuterium or "heavy hydrogen" is the isotope of hydrogen with a neutron in its nucleus, in addition to the proton which alone makes up the nucleus of normal hydrogen (the third isotope, tritium, has two neutrons plus a proton). Deuterium is found on Earth in the proportion of around 14 parts per million compared

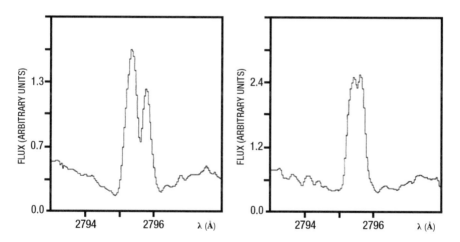

Fig. 4.9 Portions of IUE spectra of the cool stars δ Pavonis (left) and τ Ceti (right) showing emission in the k line of MgII from their chromospheres in the centres of the plots. δ Pavonis shows a deep sharp absorption within the emission line, due to interstellar Mg II whilst τ Ceti shows only a weak absorption due to self-absorption in the atmosphere of the star. We can calculate the amount of MgII in the line of sight to δ Pavonis, and hence the amount of neutral hydrogen between us and the star. This kind of measurements lets us make a 3D map of the interstellar medium around the Sun. Credits: G. Vladilo et al. 1985 Astronomy & Astrophysics. Vol 184, p. 81

with hydrogen. The key astrophysical reason to look for and measure deuterium is that theories of element production in the primordial fireball of the Big Bang predict an abundance at around this value, while theories of nuclear synthesis and destruction in stars suggest that deuterium is destroyed rather than created in stellar interiors. Indeed the nuclear astrophysicist William Fowler, who shared the Nobel Prize for his work on the synthesis of the element nuclides in general (in stars, the Big Bang, and the interstellar medium) once remarked that he would bet that deuterium would never be observed outside the Earth. Fortunately he was a better nuclear physicist than astronomer and deuterium has been observed in a variety of astronomical settings. The first detections of deuterium outside the Earth were made in the 1970s by by observing molecules in Jupiter's atmosphere in the near infrared, and in the 1970s the Copernicus satellite detected it in the interstellar medium quite locally within our Galaxy. But the major step forward in cosmic deuterium abundance measurements had to await the launch of the FUSE, the first far ultraviolet satellite. This was designed to produce high resolution spectra in the wavelength range 90.5–118.7 nm (around one third of the wavelength at which the visible transits into the UV, close to 300 nm). This covers most of the Lyman lines of atomic hydrogen. The lines of deuterium always appear close to the equivalent lines of hydrogen and it needed a spectrometer of high enough resolution to separate them well. FUSE provided this resolution, and its measurements resolved a problem revealed by the Copernicus data. The ratio of deuterium to hydrogen varies by large factors from one place to another in the interstellar medium. If all the deuterium was produced in the Big Bang, and a fraction destroyed inside stars, it should be quite uniformly mixed in space. The FUSE measurements showed that the deuterium to hydrogen ratio is much lower in the spectra of stars whose spectra show that they are in a region of interstellar gas with a high dust content. Work by astrochemists then showed that deuterium binds to the dust much more efficiently than hydrogen, so that the measured deuterium abundance in those directions is not a valid estimate of the intrinsic ratio between them, so that only the maximum value of the ratio could give a reasonable value for the primordial deuterium abundance. This is currently set at around 25 parts per million.

4.3.7 Galaxies in the UV

When we observe nearby galaxies in the UV we pick out their hottest components, the bright, blue, young, massive stars. The most striking view of this kind is seen in Fig. 4.10 which shows a GALEX image of the Andromeda

Galaxy, M31, which we have already seen in the visible and in the middle infrared. The UV emission has been transformed to blue in Fig. 4.10 so that it can be seen, which intuitively shows us the hot component stars. These are mostly picking out the spiral structure, and we know that hot, blue stars are forming continually in the spiral arms of galaxies such as the Milky Way and Andromeda. But in Andromeda the spirals are wound around to such an extente that they look almost circular. By detecting the UV we can see this with clarity, but surprisingly when we juxtapose this image to the image of the same galaxy taken in the mid infrared with the Spitzer satélite, in Fig. 3.19, we find quite similar structure. This is because the infrared image is picking out the warmest part of the dust in the interstellar medium, which is found in the gas surrounding the hot stars. The large scale coincidence which we see would give way to more complex structures if we had images with the resolution to pick out the individual star clusters and their shrouding dust clouds. If we compare these two images with Fig. 1.7, taken in the optical wavelength range, this image is very different because it picks up a much wider range of stars, from the hottest most massive to the much cooler less massive stars, but it also registers some emission from the ionised gas which produces a diffuse shimmer effect around much of the disc. These three

Fig. 4.10 UV image of the Andromeda Galaxy from GALEX. Credit NASA/ JPL Caltech

figures are in themselves an epitome of multi-messenger astronomy in its-multi-wavelength style, of a galaxy. Widening the wavelength range to radio, on the long side, and to X-rays on the short side will probe even more types of objects and processes in galaxies. This is illustrated directly in Chap. 12 where Andromeda imaged in different wavelength ranges is displayed in Figs. 12.21, 12.22, and 12.23.

4.3.8 Supernova 1987A

This supernova, in the nearby small galaxy, the Large Magellanic Cloud, is the nearest known supernova to Earth observable during the epoch of modern observational techniques. It also appears in this book in the context of the beginnings of neutrino astronomy. It was observed by IUE within 14 h of its reported discovery. The IUE spectra were used to show that just before the explosion the star had been a blue supergiant, which was not expected by the supernova theorists. The Hubble space telescope photographs of the zone previous to the explosion revealed an expanding nebula around the progenitor star, which was the result of a massive stellar wind. This nebula could then be studied after the explosion by IUE, whose spectra detected a large abundance of nitrogen, one of the key elements in the nucleosynthesis which goes on inside really massive stars. The inference was then made that the star had been a red supergiant but had lost a significant fraction of its mass due to the stellar wind, which converted it into a blue supergiant before it exploded. The results gave astronomers many insights into the physics of stars just before and just after a supernova explosion. Figure 4.11 is a composite image of SN 1987a from images in the visible+near UV, in X-rays and in the millimetre wavelength radio regime.

4.3.9 The Lyman-α Forest, Quasars, and the Reionisation of the Universe

In 1970 US astronomer Roger Lynds observed the spectrum of a quasar, 3C 05.34, an object at a redshift of 2.9, which means that we are seeing its light emitted when the universe was just over a quarter of its present age. Quasars had been discovered and tentatively identified during the previous decade, as we can read in the chapter on radioastronomy, and Lynds was working on one of the earliest high resolution spectra, which meant that he could see

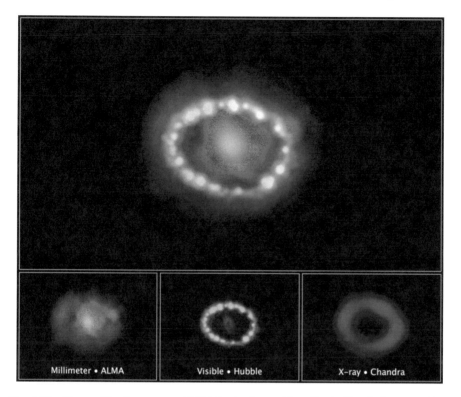

Millimeter • ALMA | Visible • Hubble | X–ray • Chandra

Fig. 4.11 Upper: The Supernova 1987A in the Large Magellanic Cloud photographed by the Hubble Space Telescope, overlaid by millimetre wave data from ALMA and X-ray data from the Chandra satellite observatory. The circle of blobs is due to matter in expansion expelled from the progenitor star 20,000 years before it exploded, now being overtaken by the more rapidly expanding supernova shell, picked out in X-rays. The emission detected by ALMA is interstellar dust emit produced by the supernova, which emits at submillimetre wavelengths. Lower. The three separate images which were used; left the Chandra X-ray image coded in blue, centre, the Hubble optical image in green, and right the ALMA submillimetre image coded in red. Credits: ALMA (ESO/NAOJ/NRAO) /A. Angelich. Visible and near UV image: the NASA/ESA Hubble Space Telescope. X-Ray image: The NASA Chandra X-Ray Observatory

hundreds of spectral lines to identify. He noticed a considerable concentration of absorption lines whose identity was hard to establish, but from their general wavelength range he concluded tentatively that they were produced by hydrogen, moving at a range of velocities and so each absorber must be an object moving at a different discrete velocity with respect to us. The velocity of recession of the quasar itself had been measured using lines due to different elements such as those of quintuply ionised oxygen (0VI) and also the strongest line of hydrogen, Lyman-α, seen in emission from the quasar. The concentration of unknown absorbers was at shorter wavelengths, bluer, than Lyman-α, but close to it. At that time there was still a controversy between the supporters of the Steady State theory of cosmology and those who favoured the "Big Bang" expanding universe. According to the latter the line wavelengths of the quasars were due to cosmic expansion, and their velocities were directly measurable by their redshifts, the change in wavelength of their spectral lines compared with those emitted by the same atoms in the laboratory. The redshifts of the quasars were very large compared with those of known objects such as galaxies, which put them at huge distances, and made their power output, their luminosities very high, too high for conventional ideas about energy production. In the Big Bang models, the multiple Lyman-α absorption lines seen by Lynds, and later verified in other quasars by a number of observers, had a natural explanation as due to clouds of hydrogen lying along the line of sight between us and a given quasar, a suggestion made first by the Dutch radioastronomer Jan Oort. Each cloud lies at a different distance from us, and so is receding at a different velocity, and each cloud gives a separate Lyman-α absorption line. But the apparently unattainable power of the quasars if they were indeed at cosmological distances strengthened the conviction of the Steady State supporters, who looked for other explanations. The eventual clincher about the nature of the quasars themselves was the gradual revelation that at the nuclei of galaxies the matter tends to fall into a deep gravitational potential well, and this has given rise to the presence of supermassive black holes. The efficiency of the production of radiation as matter falls into a black hole is some hundred times higher than that of conversion of hydrogen to helium by nuclear reactions in stars. So the immensely powerful highly compact sources of radiation which are the quasars, have been fully and reliably identified as supermassive black holes at the centres of galaxies.

The picture of the Lyman-α forest itself was gradually cleared and one of the key steps was the ability of UV astronomers to obtain spectra of quasars much nearer to us. As these are at low redshifts, their Lyman-α emission lines are in the ultraviolet, and their Lyman-α forest absorption lines are further into the ultraviolet. Figure 4.12a is a schematic to show how the Lyman-α forest is formed, and Fig. 4.12b is a spectrum of a relatively nearby quasar showing its own Lyman-a emission and the small number of intervening cloud absorptions.

As more and more quasar spectra were observed over a wide range of redshifts, it emerged that the number of these absorbers grows with increasing

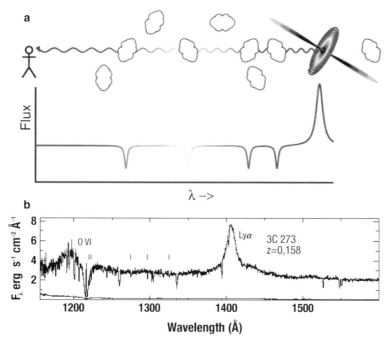

Fig. 4.12 (a) Schematic showing how the Lyman-α forest is produced. Light from a quasar, i.e. from the surroundings of a supermassive black hole at the centre of a galaxy, is emitted and includes the Lyman-α line shown as the red peak in emission. Clouds of neutral hydrogen between the observer and the quasar each causes a separate absorption line. Their wavelengths show their cosmic expansion velocities, which are smaller, relative to us, the nearer the cloud. This is represented by showing the lines increasingly bluer for the closer clouds. Credit: Ned Wright, used with permission. (b) The spectrum of 3C273, one of the nearest quasars, taken with the GHRS spectrograph on the Hubble Space Telescope. The Lyman-a emission from the quasar is intense. The intervening "forest" lines, positions marked with vertical lines, are few and faint, as the object is very close in cosmological terms. As the recession velocity (redshift) of 3C 273 is quite small, the forest lines are not strongly redshifted, and are all in the UV. Credits: NASA STScI/ J. Bahcall et al. 1993 Astrophysical Journal Supplement Series, Vol. 81. P. 1

redshift, which strongly supports the idea that we are seeing clouds along the line of sight, because of the general argument that the further away we reach when observing a quasar, the more clouds will be penetrated by our line of sight. In nearby quasars the UV spectrograph GHRS was used for some of the earliest observations of the local Lyman-α forest and their relative sparsity over short distances helped to separate them clearly and thus to be able to measure the properties: density, temperature, total mass, and metallicity

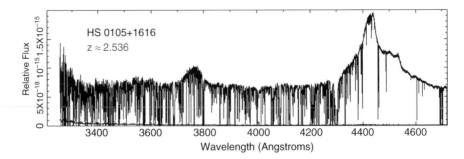

Fig. 4.13 High resolution spectrum of the quasar HS 0105+1619, taken with the HIRES spectrograph on one of the Keck 10.4 m telescopes. The vast majority of the absorption lines are due to absorption by atomic hydrogen in individual clouds between the quasar and the observer. The larger the redshift of a quasar, the more potential clouds along the line of sight. Credit Astrophysical Journal Vol. 552, p. 718, 2001 Figure 1, O'Meara et al

(proportions of different elements) using Lyman-α and absorption lines of other elements. The fact that each cloud has its individual velocity (redshift) means that we can tag the imprint of each cloud on the spectrum, with all its spectral lines, separately (Fig. 4.13).

At higher redshift the Lyman-α forest is shifted into the visible and even into the near infrared, and can be observed with the largest telescopes on the ground. We can see an example of this in Fig. 4.13, the spectrum of a quasar at redshift 2.536 taken on one of the two Keck 10.4 m telescopes, (Hawaii). At redshift 6 and beyond, the individual Lyman-α absorption lines blend into each other and form a continuous wide trough. This, the Gunn-Peterson trough, is due to the increasing fraction of neutral, unionised hydrogen as we go back to a range of epochs when the expanding universe as a whole had cooled down, leaving almost all the hydrogen neutral, but before there were enough stars and galaxies to ionise most of the hydrogen in the intergalactic medium (Fig. 4.14).

The Lyman-a forest offers astronomy a unique tool to probe structure on large scales. Its study enables astronomers one means of comparing the predictions of cosmological model simulations with observations. Because it probes in velocity, the Hubble-Lemaître expansion gives us the means to make three-dimensional maps of the intergalactic medium,

The "fog" of neutral hydrogen detected in the Gunn-Peterson trough makes it difficult to see back further towards the Big Bang using ultraviolet, optical, and near infrared radiation. Going back further shortly after the Big Bang, after the formation of protons and electrons, the universe as a whole was too

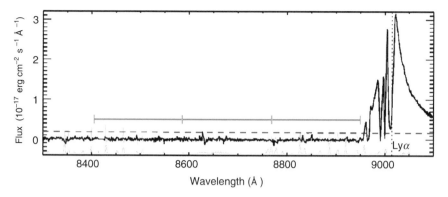

Fig. 4.14 The spectrum of a quasar at a redshift of 6.4 emitting when the universe had less than one seventh of its present age. The flat zone to the blue of 8875A represents the universe essentially opaque to ionising radiation, before the stars and galaxies "reionized it". This effect was discovered by Gunn and Peterson. Credits: T. Goto et al. 2011 Monthly Notices of the Royal Astronomical Society, Letters, Vol. 415. P.1

hot for neutral hydrogen to form, so at that time the whole universe was ionised, but as it expanded and cooled the electrons and protons combined and the hydrogen became neutral, atomic hydrogen. This occurred some 375,000 years after the Big Bang, and it is the radiation from this, the "recombination epoch" that we detect today in microwaves as the "Cosmic Background Radiation". Then came a period of gradual cooling, and the hydrogen remained neutral, until the first stars and galaxies formed. During this epoch, when the universe was around half a billion years old, we cannot trace the phenomena in the universe via the UV, visible, or the IR. The radiation from these new objects then ionised the hydrogen (its name, the re-ionisation epoch, reminds us that before 375,000 years the universe was also ionised) leaving only the denser clouds still neutral, and it is from this epoch, until the present day, which we can explore using the Lyman-α forest. Figure 4.15 is a well-known diagram which aims at summarising the evolutionary history of the universe, taking into special account the initial cooling, the formation of hydrogen, then neutral hydrogen, as the main constituent, the subsequent condensation into stars and galaxies, which in their turn reionised the universe, making it sufficiently transparent to detect individual sources. The epoch between the first combination of protons and electrons into hydrogen, releasing the energy of the cosmic microwave background radiation, and the reionisation caused by the stars and galaxies, was termed by theorist Martin Rees the "dark ages". We introduce this figure here as a taster. In the chapter

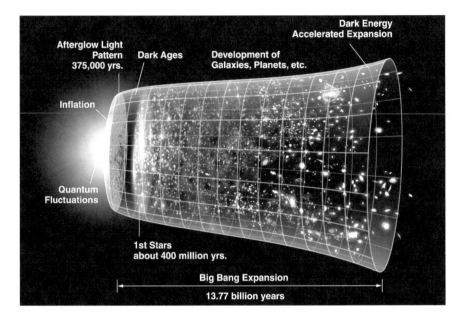

Fig. 4.15 Highly schematic timeline of the currently favoured "Big Bang" cosmology. After the initial period of very rapid expansion under cosmic inflation, the universe cooled until its protons and electrons joined to form neutral hydrogen. The radiation separated from the gas, forming what we now measure as the cosmic microwave background (375,000 years after the Big Bang). There followed the "dark ages" in which light could not penetrate the neutral hydrogen, but during this time the atoms formed clouds which condensed into stars and galaxies. These reionised the gas, allowing detectable radiation to escape from individual objects. We see those galaxies and quasars which are nearer to us in time and space than the end of the dark ages. The Gunn-Peterson effect detects the stage of emergence from the dark ages. The Lyman-a forest allows us to explore the structure of the universe since that epoch. Credits: NASA/WMAP Science Team/Cherkash

on cosmology we will look particularly at the earliest part of this timeline, when the nuclei of the atoms were being formed, in immensely hot and dense surroundings.

Further Reading

Book Barstow, Martin A., Holberg, Jay B., Extreme Ultraviolet Astronomy (book) Cambridge University Press, 2007. https://www.cambridge.org/es/academic/subjects/physics/astrophysics/extreme-ultraviolet-astronomy?format=PB#contentsTabAnchor

Review article: Linsky, Jeffrey L. "UV astronomy throughout the ages" Astrophysics and Space Science Vol. 363 2018 Springer Link. https://link.springer.com/article/10.1007/s10509-018-3319-9#citeas

Book based on a conference, Gomez de Castro, A. I. (Editor) Ultraviolet Astronomy and the Quest for the Origin of Life. Elsevier, November 2020.

5

X-Ray Astronomy

5.1 First Steps in X-Ray Astronomy: Rockets, Balloons, and Early Satellites

It may seem ironic that X-rays, the part of the electromagnetic radiation which we most commonly use to penetrate and image the human body, are not able to reach the ground and therefore X-ray astronomy has to be performed from above the atmosphere. A technique for observing X-rays from objects beyond our atmosphere using rockets was developed by Edward Hulbert as early as 1929, but the first practical observations were made by US scientist Herbert Friedman in 1949. He used detectors mounted on captured German V2 rockets from the Second World War to detect X-rays from the Sun. In the 1950s and 1960s the UK launched a series of observing experiments using X-ray imagers in Skylark rockets to produce high quality X-ray pictures of the Sun.

Since those early days, there have been a number of increasingly sensitive and powerful solar X-ray observatories in space, and we will look at some of their results in a later part of this chapter. In 1962 an American team, Giacconi, Gursky, Rossi and Paolini using a detector on an Aerobee rocket, (Fig. 5.1b) detected X-rays from the Sun and the Moon, but more significantly found fainter X-rays from over the whole sky, and in particular a stronger source towards the constellation of Scorpius, which was called Scorpius X–1. This marked the start of X-ray astronomy, and later Giacconi received the Nobel prize for this and his extensive subsequent work in the field.

© Springer Nature Switzerland AG 2021
J. E. Beckman, *Multimessenger Astronomy*, Astronomers' Universe,
https://doi.org/10.1007/978-3-030-68372-6_5

a

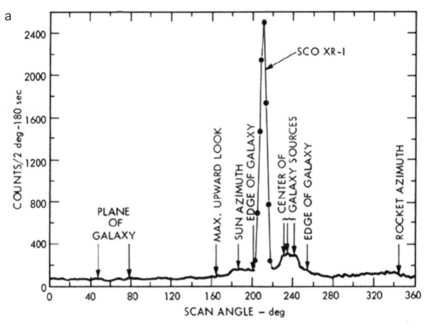

b

During the next few years astronomers used balloon-launched telescopes to prolong their X-ray observations, while Giacconi and collaborators worked on a project to develop an X-ray observing satellite. This resulted in the launch of Uhuru, the first orbiting X-ray observatory, in 1970, which observed photons in the 2–20 keV range (Fig. 5.2).

This ground-breaking project made very significant advances. It detected the sources Cen X-3, Vela X-1 and Her X-1, in the constellations of Centaurus, Vela, and Hercules, respectively. These sources are all powered by the accretion of material from binary companions onto rotating neutron stars (pulsars), and X-ray observations have been of critical importance in our understanding of these systems. It also detected Cygnus X-1, which was the first candidate for a black hole in the stellar mass range, as well as many important extragalactic astrophysical sources. A series of catalogues was published, with the last of these, 4U, containing 339 objects, covering the whole sky in the 2–6 keV band. Uhuru was active from December 1970 to March 1973, and its satellite finally fell into the atmosphere on April 5th 1979.

The interest in X-ray astronomy grew rapidly from 1970 onwards, and the result was a series of X-ray observing satellites, more than I want to describe here, so I will describe a selection of those that I find most interesting. NASA was the most prolific producer, as may be expected from the most important and well funded space agency, and launched a series of HEAO (High Energy Astronomical Observatory) Satellites, of which the most productive was HEAO-3, later named Einstein, which had the first real X-ray imager, observing between 0.1 and 6 keV energies, with an angular resolution of order 10 arcseconds. ESA launched a general X-ray observatory, EXOSAT, which was operational only between 1983 and 1986. The Japanese Space Agency, JAXA, specialised in solar X-ray astronomy. It flew Hinotori to detect solar flares in X-rays between 1981 and 1991, Yokoh, which observed solar X-rays across the full X-ray energy range, was launched in 1991 and ended its mission in 2005, while its successor Hinode started operations in 2006 and is still active. As well as an X-ray imager Hinode also has an optical imager and an extreme

---

Fig. 5.1 (a) X-ray scan across the sky taken during three minutes of a rocket flight in 1967. The strongest source is Sco X-1, discovered by Riccardo Giacconi in a rocket flight spectrum taken in 1962. The general emission observed as the scan crossed the plane of the galaxy is also seen above the background level. Credit: HEASARC/NASA-GSFC. (**b**) Riccardo Giacconi, Italian-American astronomer "the father of X-ray astronomy" Director General of the European Southern Observatory (ESO) 1993-1999, Nobel Laureate in Physics, 2002. Credit: ESO

Fig. 5.2 X-ray satellite SAS-1, also known as X-ray Explorer, (and later renamed UHURU) in a preparatory phase at Goddard Space Flight Centre. Project manager Marjorie Townsend and astronomer Bruno Rossi discussing details. Credit: NASA GSFC

ultraviolet imaging spectrograph, at wavelengths bordering on soft X-rays. The Japanese ISAS agency launched a general X-ray astronomical observatory, Tenma, in 1983, which operated for only 2 years, and was followed by Ginga, launched in 1987 and operational till 1991. The Soviet space agency flew the Granat satellite, with a hard X-ray telescope and also a gamma-ray telescope, acting as an all-sky observatory, between 1989 and 1998. The Italian astronomical community studied variable astronomical X-ray sources with its BeppoSAX satellite, launched in 1996 and operational for 7 years. The most interesting and productive missions, in my view, include the German Space Agency's ROSAT (produced in collaboration with the UK and NASA) which flew from 1990 to 1999, NASA's Chandra observatory, launched in 1999 and still operational, and ESA's XMM-Newton satellite, also launched in 1999 and also operational today. Finally in this list we will see a result from the Spektr-RG X-ray all-sky observatory, produced by the Russian Space Research Institute and the German Max Planck Institute for Extraterrestrial Research, and launched in 2019. But before I describe what is being observed, it is important to describe how the observations are made, because X-ray astronomical techniques mark a major departure from conventional optics as used in the visible, the infrared, and the ultraviolet.

5.2   X-Ray Detectors and Telescopes

The early X-ray astronomical missions used proportional counters to make their detections. A proportional counter is a tube filled with an inert gas at low pressure to which an electric field is applied by a high voltage between two electrodes. X-rays can enter the tube through a window; they ionise the gas and the resulting electrons are accelerated by the voltage, colliding with neutral atoms in the gas, producing more ionisation. The result is an electric current whose total charge is proportional to the energy of the incoming X-ray photon. The principle is the same as that of the famous Geiger-Müller counter used to detect radioactivity in general, but in the latter the voltage is very high, so each particle is detected as a burst of electrons, with no attempt to measure the energy of the particle. At the lower voltage used in a proportional counter, the net charge produced is proportional to the energy of the X-ray photon, which can thus be measured. Proportional counters were used in the first two decades of X-ray astronomy, from rockets and balloons, as well as in the earliest satellites. Figure 5.3 shows the basic principle of a proportional counter.

As with any scientific instrument one of the most important aspects is not only to make a detection but to reject detections of other types of particles. As well as using physical shielding this is done within the electronics of the detector, as different particles give different time responses, and particles with energies outside the detection band can also be rejected. Another type of detectors used for X-rays on astronomical satellites is a scintillation detector. The heart of such a detector is a crystal of material chosen so that X-rays in the range to be observed can release an electron from an atom with considerable energy. This electron then collides with further atoms, exciting their electrons into higher states, from which photons are emitted at visible or near visible wavelengths. These photons can then be detected by standard techniques for the visible.

Work on X-ray detector technology has progressed steadily. An element introduced in the late 1970s was the micro-channel plate. This is a disc of glass, approximately 2 mm thick, with hundreds of fine straight tubes with diameters between 5 and 10 μm between the two faces of the disc. The front surface of the disc is coated with a photoelectric metal. An impinging X-ray releases an electron from this surface, and an applied voltage across the disc accelerates the electron along a tube, without any attempt to limit its direction. As it collides with the walls more electrons are released, and the result is a strong signal of at least 1000 electrons per detected X-ray photon. These plates can be used in series to achieve amplification factors between a million and a hundred million. The plates have to be shielded in all directions except along the beam of incoming astronomical X-rays. From the end of the 1970s CCD's of essentially the same types as are used in optical cameras, were shown to make good X-ray detectors. They function well in X-rays, and it is not difficult to shield them from normal light by simple encapsulation in thin sheets of material opaque to light but transparent to X-rays. So modern X-ray telescopes in space use essentially the same detection techniques as modern optical telescopes: CCD's or less commonly photon counting systems based on microchannel plates.

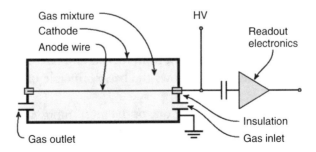

Fig. 5.3 Diagram of a proportional counter for X-ray detection used in the first generations of X-ray satellites. Credits: A. Winkler, E. Brücken et al./U. Helsinki

5.2.1 Grazing Incidence Optics: Einstein and Rosat X-Ray Satellites

The main difference between an X-ray telescope and an optical telescope is the way the X-rays are focused to make an image. When an X-ray impinges on any solid surface it penetrates until it interacts with the atoms in the solid. This means that a normal mirror will not reflect a significant fraction of the X-rays, and a normal lens will not bring them to a focus. The way round this problem is to use surfaces coated with a heavy metal, such as gold, and to use reflection at "grazing incidence", when the ray makes a very small angle with the reflecting surface. An illustration of this principle is given in Fig. 5.4. It is Type I of three types of optics of this kind devised for X-ray work by Hans Wolter, in 1952. All X-ray satellite telescopes use Wolter's designs.

These focusing instruments have shown themselves capable of matching all but the most powerful optical and infrared instruments in terms of angular resolution. The first focusing telescope using X-rays was flown in NASA's 2nd High Energy Astronomy Observatory (after UHURU); this was first called HEAO-2 and was then renamed Einstein, and observed from 1978 to 1981. The telescope fed four detectors, the HRI, a high resolution imaging camera, in the range 0.15 keV–3 kEV, and imaging proportional counter from 0.4 to 4 keV, a Solid state spectrometer, from 0.5 to 4.5 keV, and a Bragg Focal Plane Crystal spectrometer.

The European contribution to X-ray astronomy in the 1990s was ROSAT (ROentgen SATellite), developed in Germany with contributions from the

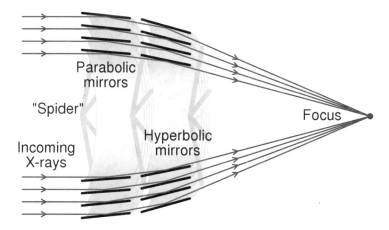

Fig. 5.4 Schematic principle of an X-ray telescope using grazing incidence optics. Credits: Cmglee, CC BY-SA 4.0 (https://creativecommons.org/licenses/by-sa/4.0), via Wikimedia Commons

US and the UK, which was launched in 1990 and flew until 1999. It was not a large satellite but contained several instruments and made an important all-sky survey. Figure 5.5b shows the optics of the satellite, and Fig. 5.5a is a diagram of the X-ray telescope and its focal plane.

5.2.2   The Chandra X-Ray Observatory

A larger and more powerful X-ray telescope was launched in 1999, NASA's Chandra X-ray Observatory, (named after the Nobel laureate Indian-US astrophysicist Chandrasekhar) which is still producing observations today. Chandra was three orders of magnitude more sensitive than Einstein, and 10 orders of magnitude more sensitive than the instrument which detected the

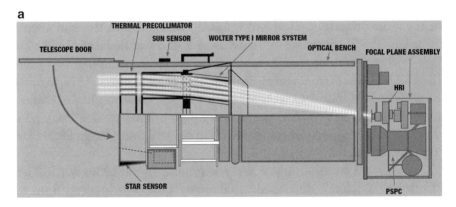

Fig. 5.5 (a) Schematic of ROSAT showing the Wolter optical system of nested grazing incidence mirrors to focus the X-rays on the instruments in the focal plane. Credit: Max Planck Institut für Extraterrestrische Physik (MPE/IAC-UC3). (b) Internal Optics of the ROSAT X-ray satellite Credits: DLR/NASA

first extra-solar X-ray source, Sco X-1. Figure 5.6 is a diagram of the satellite as deployed on orbit The grazing incidence optics give a very long focal length, so the distance from the mirrors to the focal plane is 10 m. The overall optical scheme is similar to that of ROSAT but scaled up, which gives greater collecting power for the X-rays and also higher angular resolution, reaching 1 arcsecond on the sky (Fig. 5.7).

5.2.3   XMM-Newton

While the Chandra satellite observatory was built for high quality high resolution imaging the XMM-Newton X-ray satellite built by ESA, the European Space Agency, launched in 1999, was the first X-ray mission to carry out spectroscopy. It comprises three separate telescopes, each with 58 nested coaxial grazing-incidence mirrors, and three instruments: EPIC, the European Photon Imaging Camera (one at the prime focus of each telescope); RGS, the Reflection Grating Spectrometer, fed by two of the three telescopes; OM, the optical monitor, a sensitive visible and UV telescope which observes the same region as the X-ray telescopes to give knowledge about the nature of the sources detected in X-rays. The X-ray beam from the telescope is distributed 53–47% between the EPIC and RGS The telescope has a similar basic design to those of Chandra and Einstein, but the spectrometer RGS has a complex structure, to combine focusing with wavelength dispersion using grazing incidence. There are two grating arrays in RGS, each with 182 identical gratings; a view of one of these arrays is shown in Fig. 5.8.

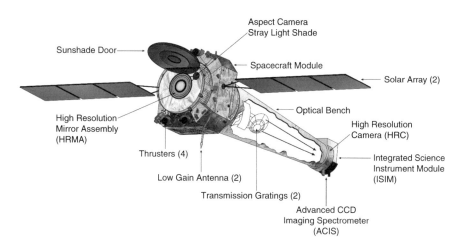

Fig. 5.6 Schematic diagram of Chandra X-ray satellite deployed on orbit. Credits NGST and NASA/CXC/R. Vaughan

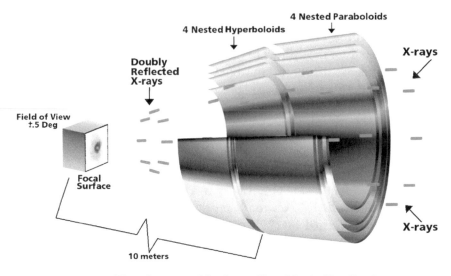

4 Nested Hyperboloids

4 Nested Paraboloids

**Doubly
Reflected
X-rays**

X-rays

**Field of View
±.5 Deg**

**Focal
Surface**

X-rays

10 meters

Mirror elements are 0.8 m long and from 0.6 m to 1.2 m diameter

Fig. 5.7   The optical scheme of Chandra. Credits: NASA/CXC/D. Berry

Fig. 5.8   One of the grating arrays on the XMM-Newton satellite. Credit: ESA

5.3   What We Learn from X-Ray Astronomy

5.3.1   What Kinds of Objects Produce X-Rays

5.3.1.1   Solar X-Rays

Since the pioneering days of the 1960s various countries including the US (NASA), Western Europe (ESA), Russia, Japan, India and China, have launched satellites or other probes to investigate solar phenomena. We cannot cover many of these individual missions, but will combine a few historic points to show the science which has emerged. The first ambitious solar X-ray image flew on NASA'a Skylab in 1973–4. It was the S-054 X-ray camera mounted on the multi-use "Apollo Telescope Mount", and showed for the first time the structure of the solar corona. An image from this is shown in Fig. 5.9.

The reason that X-rays image the corona is its very high temperature, which ranges up to 2 million K, and with even higher temperatures in individual outbursts. This high temperature had surprised the astronomers in the 1930s and 1940s when it was discovered, because the photosphere below it, which emits the visible light we see, is much cooler, at around 5700 K, and the detailed mechanisms which supply the energy to the corona are still a subject of active investigation. Measurements in X-rays gave unquestionable values of the coronal temperature, but as we can see, even in the early image in Fig. 5.9, they also revealed the complex coronal structure. The darker zones are "holes" where the large scale magnetic field has guided the upflowing energy away and into the hotter zones of the corona which are represented as yellow-white, while the cooler zones are orange-red. The resolution is moderate but even so we can see structure on quite small scales. As technology improved and the instrument size increased more precise imaging became possible, with a range and variety of X-ray wavelengths each probing a different range of heights within the corona. As an example we can take Fig. 5.10 which is a composite of X-ray and ultraviolet images. The regions of greatest activity are in blue, as taken in 2015 by NASA's NuSTAR (Nuclear Spectroscopic Telescope Array) satellite, less active regions in green from JAXA's (the Japanese Space Agency's) HINODE satellite, and far ultraviolet emission (very close to the X-ray bands) from NASA's solar dynamics observatory, SDO.

As the X-ray spectrum uniquely pinpoints the solar corona, static and dynamic X-ray imaging, combined with studies of the photosphere and the chromosphere in the visible and ultraviolet, allow us to understand the

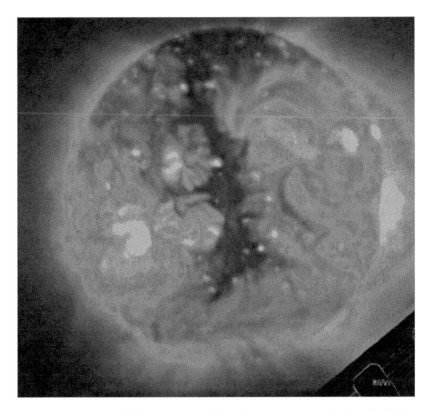

Fig. 5.9 X-ray Image of the Sun taken from the camera on the Apollo Telescope Mount on Skylab in 1973 Credits: NASA/AS&E/NSSDC

processes by which energy is transferred from the sun's surface outwards, and eventually to the solar wind which impinges on the Earth's outer atmosphere. Of particular practical interest are the solar flares, events in which a sudden release of magnetic energy drives out a pulse of high speed sub-atomic particles. Their impact on the Earth's atmosphere can have a drastic effect on terrestrial and space communications, and even potentially on the health of airline crew if a particularly intense pulse penetrates into the stratosphere. NASA has implemented a series of satellites named GOES which are in geostationary orbits, and which, as well as making large scale observations of Earth's weather, monitor the Sun in X-rays, in order to offer a warning of "bad weather" in space. The X-rays from a flare take the same time as visible light to reach us, some 8.5 min, while the particles take several hours, so air crews and astronauts can take a minimum of precautionary measures.

Figure 5.11 shows a set of images taken with GOES 17 through a series of filters between the far ultraviolet and the soft X-ray spectrum in which a flare

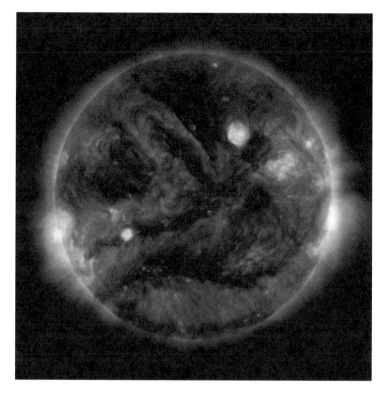

Fig. 5.10 The Sun. A composite of X-ray and ultraviolet images. The regions of great-est activity are in blue, as taken in 2015 by NASA's NuSTAR (Nuclear Spectroscopic Telescope Array) satellite, less active regions in green from JAXA's (the Japanese Space Agency's) HINODE satellite, and far ultraviolet emission (very close to the X-ray bands) from NASA's solar dynamics observatory, SDO. Credits: JAXA/NASA

is imaged in the top left quadrant of the solar disc. Through different filters the emission from different elements and their ions are imaged, which allows the experts to build up a picture, over time, of the processes within the corona, including flares. At optical wavelengths the imaging of a flare in the lower solar atmosphere, the photosphere and the corona, can be spectacular, as we show in Fig. 5.12, taken from the Solar Dynamics Observatory.

But the principal contribution of X-ray observations has been to quantify the processes at work in the high temperature region above this, i.e. in the corona. In Fig. 5.13 you can see X-ray images of the sun around an 11 year cycle of activity, taken with the Japanese-American YOKOH satellite. During solar máximum not only are there many flares, but the whole of the corona emits more strongly in X-rays, because it is being heated more strongly from below.

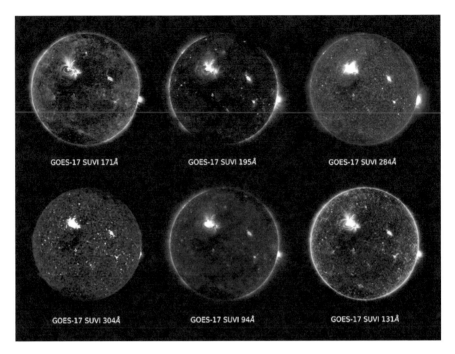

Fig. 5.11 A set of images of the Sun taken with GOES 17 through a series of filters between the far ultraviolet and the soft X-ray spectrum in which a flare is imaged in the top left quadrant of the solar disc. Credits: NASA/WMO

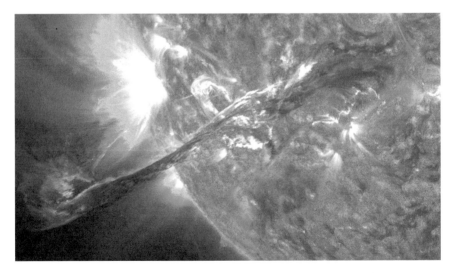

Fig. 5.12 A solar flare on August 31st 2012 imaged in the Hα line from the Solar Dynamics Satellite Observatory which observed in the visible and ultraviolet range, into the softest X-ray region. Credits: NASA/GSFC/SDO

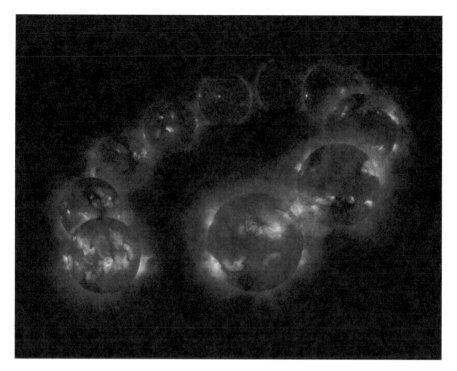

Fig. 5.13 X-ray images of the Sun during an 11 year sunspot cycle, starting in August 1991 at sunspot máximum taken by the YOKOH X-ray satellite. The more active regions, the flares, and the coronal holes are all present in these images, and we can see that at solar mínimum half way round the cycle the whole corona is much less brilliant in X-rays than at the two solar máxima at both ends of the cycle. Credit: JAXA

5.3.1.2 X-Rays from the Local Interstellar Medium

The volume of space around the Sun in the plane of the Milky Way is filled with hydrogen gas, but in clouds with very different temperatures. Radioastronomers detect the cool neutral atomic hydrogen, at temperatures in the region up to around 100 K (lower than −200 °C), optical astronomers detect the warm ionised hydrogen in clouds at temperatures of order 10,000 K, far infrared astronomers detect the cold clouds of molecular hydrogen at temperatures below 20 K (−250 °C) but it was not until X-ray astronomers applied their techniques that we were sure of the existence of another phase in the local interstellar medium: a hot phase at temperatures over well over 1,000,000 K. This is directly detected because it emits X-rays. An example is given in Fig. 5.14 of the near surface of a superbubble whose boundary is only a few hundred light years away from us. The structure was first detected in

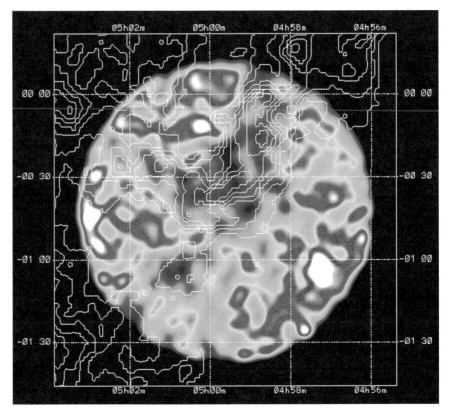

Fig. 5.14 A portion of the near surface of the Orion-Eridanus Superbubble, observed in soft X-rays by the ROSAT X-ray observatory—The bubble has a total diameter some 1200 light years and contains gas heated by the stellar winds and supernovae in the Orion OB1 association of massive young stars. Blue areas are zones with less X-ray brightness and the dark blue zone in the upper middle coincides with a filament of dust and gas which emits in the far infrared (white contours). The filament is between us and the bubble, which is heating the dust and causing the strong infrared emission. Credits NASA/DLR/David Burrows/Zhiyu Guo

atomic hydrogen in the 1970s, but the X-ray measurements showed that a large fraction of it is filled with million degree gas.

This conclusion is true for a large part of the interstellar medium within a few thousand light years of the Sun; the hot gas occupies over 90% of the local volume. The heat is supplied by the young massive stars which feed huge energy fluxes into the medium in the form of winds and UV radiation while they are "alive" and via supernova explosions at their deaths. Although the hot gas is all-pervasive, the mass fraction of cool gas, neutral hydrogen at under 100 K, is about half of the total, with the hot gas at 1,000,000K at somewhat less than half. Remember that the hot gas is 10,000 times hotter than the cool gas, but if they are in pressure equilibrium, or nearly so, the density of the cool gas must be around 10,000 times greater: a very small cloud of cool gas will have as much mass as a very big cloud of hot gas. This makes the interstellar medium very inhomogeneous, and very porous for the escape of ionising radiation from the stars which produce it, as the hot ionised gas does not absorb ionising radiation and lets is pass, as if it were transparent. Another example of the inhomogeneity is shown in Fig. 5.15, which is a composite mosaic X-ray image of a supernova remnant the Cygnus loop, taken with

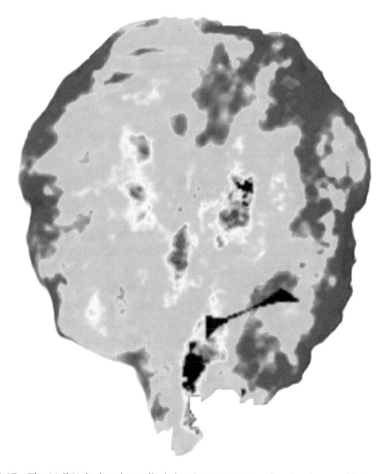

Fig. 5.15 The Veil Nebula, also called the Cygnus Loop, a local volume of hot gas produced by a supernova which exploded 10,000 years ago, imaged in X-rays with the position sensitive proportional counter on ROSAT. Credits NASA/ DLR/N. Levenson/S. Snowden

ROSAT, which we have seen in visible light and will see again at other wavelengths.

The third hydrogen component, molecular hydrogen, is found locally in only very small cold clouds embedded in the cool clouds and protected by their gas and dust from external heating and ionisation.

5.3.1.3   X-Rays from Stars

The earliest look at X-ray emission from a range of stellar types was carried out by the Einstein satellite, and reported in 1981. This found detectable soft X-rays from stars in all parts of the Hertzsprung-Russell diagram, with stronger emission coming from the earlier types, O and B stars which are the most massive and hottest, and generally lower levels from later cooler stars. However, the X-ray luminosities of the later type stars were found to be independent of their surface temperature and their surface gravity (dwarf stars have higher surface gravities than giants because the latter have inflated outer layers), showing a wide scatter. Also many late type stars show vary variable X-ray emission. There is a small range of stellar types from late B to mid-A, where X-ray activity is systematically weak, or absent.

5.3.2   Hot Stars, Young Stars

The high X-ray luminosity of the massive OB stars has been explained by their emission being due to shocks in the high velocity winds emanating from their atmospheres. Detailed spectral measurements by the Chandra satellite require the interaction of magnetic fields in these outflows to explain them. Another phenomenon for which X-rays give important clues is that of accretion of matter. Accretion plays an important role in the development of the stars themselves and more so of their planets, which are now understood to be formed in the discs of accreting gas and dust in the mid-planes, perpendicular to the axes of rotating stars. Accretion gives rise naturally to a disc in a rotating cloud, because the material approaches the disc from both sides at similar rates, and the net result is that collisions cancel out the momenta of the particles in a direction perpendicular to the disc. As this material comes from the contracting cloud from which the star had formed, it is spinning in the same sense and the net result is that the accreting matter settles into a spinning thin disc. The processes of collision in this accretion give rise to X-ray emission which can be diagnosed using specific spectral features. For example, X-ray lines of neon and oxygen, detected in stars with accretion discs imply high circumstellar densities predicted for these discs and can be used to measure them. Non-accreting stars with similar characteristics show much lower X-ray luminosities. In young stars of lower mass there is a high degree of X-ray variability, which is attributable to flares and other types of strong coronal activity. These X-ray flares, as well as accretion discs in more massive stars, have been studied by the Chandra satellite in the nearby stellar nursery, the Orion nebula.

5.3.3   X Ray Binaries

Perhaps the most interesting, and certainly the most exciting part of stellar X-ray astronomy is the subject of X-ray binaries. These are binary stars in which one of the stars is supplying material for accretion to the other, which may be a white dwarf, a neutron star, or a stellar mass black hole. If the donor star has a low mass, the compound object is termed a low mass X-ray binary (LMXB) and if the star has a high mass, generally greater than 8 solar masses, the object is a high mass X-ray binary (HMXB). The object detected and analysed using conventional optical spectroscopy is always the donor, as the compact star is generally too small to have a high optical luminosity, so it is easy to differentiate between an LMXB and an HMXB. The mechanisms for the X-ray emission in the two cases are generally different. For an LMXB the donor star has been pulled out of shape by the gravity of its compact companion, and it fills its Roche lobe, from where its material leaks out to form an accretion disc around the companion (see Fig. 5.16).

The X-rays are produced when the matter in the interior of the disc is accelerated into the deep gravitational well of the compact star; as it falls it releases energy as X-rays. This is not a uniform process, so there is considerably variability in the X-ray output. The compact object can be a neutron star or a black hole; the first X-ray source observed outside the Solar System, Sco X-1 has a black hole producing the X-ray infall. LMXB's were the first objects used to demonstrate the existence of stellar mass black holes. The key point is that there is an upper limit to the mass of a neutron star before it becomes unstable and collapses; a conservative upper limit for this is four solar masses. So when binaries with candidates for black holes were observed, the key was to use the optical spectrum to measure the kinematics of the low mass star, and derive the mass of its compact companion. The first confirmed stellar mass black hole, found by Casares, Charles, and Naylor in 1992, is the LMXB V404 Cygni in the constellation of Cygnus, whose visible star is a cool K dwarf, and

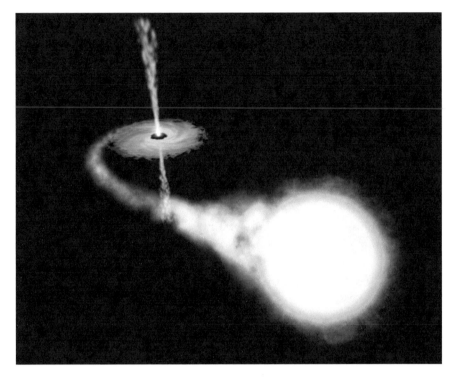

Fig. 5.16 Artist's impression of a low mass X-ray binary (LXMB), in which a low mass main sequence star is distorted by the powerful gravitational field of its compact companion. The huge tide raised fills the Roche lobe of the low mass star and leaks out to form an accretion disc around the compact star. Credit ESA/NASA/Felix Mirabel

whose compact star is a black hole of 12 solar masses. Since then some tens of these objects have been confirmed within our Galaxy.

The X-ray production mechanism for an HMXB can be the same as those for an LMXB but is more often different. The companion star is the high mass object, an O or a B star. These stars have strong stellar winds, and a fraction of the stellar wind falls onto the neutron star or the black hole, and X-rays are produced. In other objects the X-rays are simply produced in the atmosphere of the hot companion One of the best known of the HMXB objects is Cygnus X-1, one of the earliest X-ray sources detected. The compact object here is also a black hole, of 14 solar masses, and the visible companion is a hot massive O star. Cygnus X-1 emits over the full X-ray spectrum and also emits γ-rays (gamma rays) of even higher energies. It also emits strong and variable radio waves, which is reminiscent of the observable behaviour of quasars. It turns out that this is the symptom of the existence of a sharply defined relativistic (very high velocity) jet emitted along the axis of the accretion disc around the black hole. These jets are found on much larger scales emanating from the centres of active galaxies, notably quasars, which has given rise to the nickname "microquasars" applied to X-ray binaries which have them. As the X-ray binaries are within the Galaxy and susceptible to more detailed study, these microquasars can be used to investigate by analogy the physical properties of the zones around the supermassive black holes in galaxy centres which we observe as quasars. An artist's impression of one of the most observed microquasars, SS433, is given in Fig. 5.17.

5.3.4 Pulsars and Supernova Remnants

We have come across pulsars in the chapter on radioastronomy. We know that they are very dense, compact objects of a few solar masses, produced in the aftermath of a supernova explosion of a star of intermediate mass, and that they spin rapidly, emitting a narrow beam of light which we can detect as a pulse every time it crosses our line of sight. But now we also know that compact objects, white dwarfs, neutron stars, and black holes, in order of increasing density, cause X-rays to be produced as matter is pulled into their powerful gravitational fields. The Chandra satellite has carried out extensive studies of X-rays from pulsars, and in Fig. 5.18 we show its X-ray image of the Crab nebula pulsar.

This allows us to visualise the disc illuminated by the beamed energy of the pulsar as it spins in the middle, giving a really dynamic aspect, we can almost see the spin! At the same time it visualises the central jet emitted pretty well perpendicularly to the disc, from the polar regions of the spinning pulsar. In general terms X-ray astronomy is probing the high energy processes in the universe. These can be caused by very high temperatures, millions of degree plasmas, or by particles accelerated in an intense gravitational field, or in intense electromagnetic fields. The high temperatures along with intense magnetic fields are typical of the gas surrounding a supernova explosion, and impacted by its expanding shell. We can see an example of an X-ray image of the supernova remnant Cassiopeia A in Fig. 5.19a. The image has been coloured according to its energy band, with red, green, and blue coding for increasing energies. The sharp outer blast wave still expanding from the original explosion shows up in blue at the edge of the image, while the locations of the emitting elements silicon, sulphur, calcium, and iron dominate the light in the red, yellow, green, and purple, respectively. The remnant of a famous supernova explosion, seen and recorded by Tycho Brahe in 1572, can be seen in X-rays in Fig. 5.19b. This was taken with Chandra in 5 X-ray bands. The

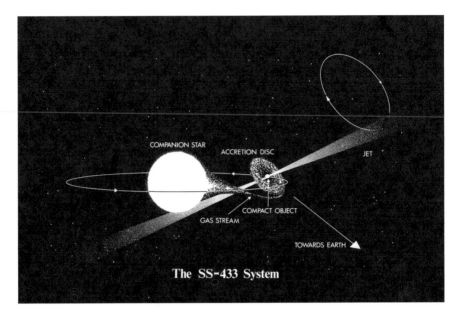

Fig. 5.17 SS-433, a high mass X-ray binary (HMXB). Gas accretes from the large star onto the compact star, which emits X-rays from around the accretion disc and also radio waves from its jet. The radiation from these objects covers a similar range to that from quasars, and they have been dubbed "microquasars" (artist's impression). Credit: European Southern Observatory (ESO)

Fig. 5.18 Crab nebula pulsar in X-rays from Chandra (blue white), visible light from Hubble (purple) and infrared from Spitzer (pink). Credits: NASA/GSFC/CXC/STScI

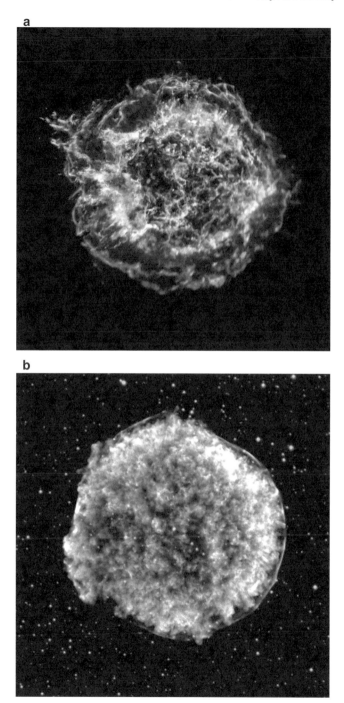

Fig. 5.19 (a) The Cassiopaeia supernova remnant imaged by Chandra. Colour coding of the X-ray bands: red 1–1.5 keV, green 1.5–2.5 keV,, blue 4.0–6.0 keV. Credits: NASA/CXC/SAO. (b) Chandra image of the remnant of Tycho's supernova. Colour coded X-ray bands: red, 0.3–1.2 keV, yellow 1.2–1.6 keV, cyan 1.6–2.26 keV navy 2.2–4.1 keV purple 4.4–6.1 keV. Total exposure time 336 h. Credits: NASA/CXC/SAO

colour of the rim shows that the highest energies are found where the expanding gas shocks the surrounding interstellar medium at the edge of the expanding remnant. The expansion is so fast that the increasing size of the spherical region is easily seen in X-ray images taken between the year 2000 and 2015. This object too was first discovered by the radioastronomers at Cambridge, and is referred to by their catalogue number 3C-10.

5.3.5   X-Ray Emission from Galaxies

From what we have already seen about stellar X-ray sources it is clear that galaxies with these types of stars, i.e. virtually all galaxies, will have X-ray emission from their stellar components. An interesting point is that treating a galaxy as a single object we would expect that its total X-ray emission should depend on the fraction of its stars in binaries, since binaries with a compact star as one of their members will make an important overall contribution. So in principle this could be used to see whether there is much variation in the fraction of binaries from galaxy to galaxy. But by far the brightest sources within galaxies are the active galactic nuclei, or AGN's. These are centres of galaxies containing supermassive black holes (SMBH's), ranging up to a few thousand million solar masses in a single object. It should not be surprising that at the centre of a galaxy there is a very strongly directed gravitational field which tends to pull the content of the galaxy towards it. However, the spin, in spiral galaxies, and the gas-like motions of the stars in ellipticals act to support the galaxy and prevent catastrophic infall to the centre. Even so there is a tendency to accumulate mass there, and most galaxies do have supermassive black holes in their centres (whether all galaxies do, or indeed whether a central black hole was necessary as a seed for each galaxy, are still open questions, under active investigation which we will not delve into here). If the matter, stars or interstellar gas, comes within range of this black hole it will be pulled in, and "on the way down" its potential energy will be converted first into kinetic energy, and then partially radiated away, principally as X-rays. It was the presence of exceptionally strong X-ray sources in galactic centres which clinched the argument that something other than a set of massive stars must be present there. Massive stars might be enough, if we include the supernovae which explode when they run out of nuclear fuel, to account for most of radiation (optical, infrared, radio) from active galactic nuclei, but the X-ray contribution is proportionately far too strong, and the total energy emitted by an AGN far too high to come from thermonuclear processes alone. These can at most convert a little under 1% of their mass into radiation, while a deep gravitational well can convert over 40% of mass to radiation, much of it in the form of X-rays. Imaging, at all wavelengths, close to the central black hole is particularly difficult. Apart from the scale, it is normally a very complex region, often filled with gas and dust, some falling in, and some being expelled. The geometry can allow us to disentangle this to some extent, and the infall is often channelled into an accretion disc perpendicular to the axis of spin of the SMBH, while the outflow is often channelled into a sharply defined on-axis jet. In Fig. 5.20 we show a Chandra X-ray image of the active galaxy Centaurus A.

This is a giant elliptical galaxy which has recently "swallowed" a smaller spiral, and the remains of the spiral are distributed around the central regions. The X-ray image, a composite of three wavebands, shows the hot extended gas in the false colour red, the jet from its central SMBH extending way outside the main galaxy, and the dust lanes nearly perpendicular to the jet, which absorb some of the X-rays. In the figure we show how, using images in different wavebands, X-rays, radio, and optical, we can build up a composite image which reveals a wide variety of different phenomena associated with this galaxy. The X-ray image also shows that there are binaries distributed throughout the galaxy.

5.3.6   X-Ray Emission from Clusters of Galaxies

X-ray astronomy made one of its greatest contributions to our knowledge of the universe by detecting and measuring the X-ray emission from clusters of galaxies. These can be considered the most massive objects in the universe, containing hundreds or up to thousands of galaxies held together by their gravitational field, and they can be millions of light years across. These clusters emit X-rays because their huge internal gravitational fields cause their intergalactic gas to try to fall towards the centre of mass of the cluster, suffering collisions which raise its temperature to millions of degrees. As it is fully ionised at these temperatures it cannot emit any spectral lines. But its thermal emission at X-ray wavelengths allows us to detect and measure it. We can artificially convert the X-ray intensity to code it in a wavelength we can see, to get an idea of the presence of this hot gas in a cluster of galaxies. The example of the cluster Abell 383 is shown in Fig. 5.21, where we can note that the hot intracluster gas permeates the cluster, densest towards the centre of the cluster as one would expect.

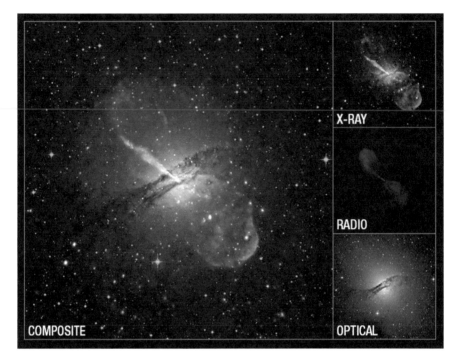

Fig. 5.20 The elliptical galaxy Centaurus A, in X-rays, optical and radio waves. To the left, the composite image, to the right the three components shown separately. This is a galaxy which amost certainly merged with a spiral, and the remains of the spiral, including its gas and dust, are circling around the nucleus. The radio image features the strong jet emanating from the central black hole, while the X-ray image shows that the power of the system in heating the gas around the whole galaxy to millions of degrees. Various Credits: /X-ray NASA, CXC, R. Kraft (CfA) et al./Radio NSF, VLA, M. Hardcastle (U. Hertfordshire)/Optical ESO/M. Rejkuba et al. /ESO-Garching

The remarkable fact is that the X-ray measurements show us that only some 4% of all the mass of the cluster is in stars, while over five times the mass in stars is in the hot intracluster medium. The rest is dark matter whose nature we still do not know, and which will feature in other chapters of this book. Using the arcs in the image, which are the gravitationally lensed images of very distant galaxies behind the cluster, the gravitational field of this dark matter can be estimated, and the blue colour shows its inferred distribution. One of the most famous results obtained thanks to combining X-ray and optical images of a cluster of galaxies is that of the "Bullet Cluster", shown in Fig. 5.22.

Fig. 5.21 X-ray image of the galaxy cluster Abell 383 The hot gas detected in X-rays is depicted in pink in this image. The optical data from the Hubble space telescope shows the galaxies in the cluster plus gravitationally lensed galaxies much further away, which show as circular arcs. The blue image shows the distribution of dark matter in the cluster inferred using these arcs. Credits: X-ray: NASA/CXC/Caltech/A. Newman et al./Tel Aviv/A. Morandi & M. Limousin; Optical: NASA/STScI; ESO/VLT/SDSS

The interpretation of this image is that two major clusters of galaxies have collided, and the hot gas from one of them has passed through the other. This must have been one of the most energetic events ever observed, but the remarkable feature comes from comparing the distribution of the hot gas, constituting most of the "normal" baryonic mass of the clusters with the total gravitational mass as measured by the lensing effects on very distant galaxies behind them, which is shown in blue. We can see a separation of the baryonic mass from the dark matter which is reckoned to make up most of the mass.

Fig. 5.22 The "Bullet Cluster" The background image of galaxies, in white and orange, is at optical wavelengths (Hubble Space Telescope and Magellan Telescope); the X-ray image of the two clusters which have collided is in red (Chandra X-ray observatory). The blue shows the distribution of dark matter inferred from the images of gravitationally lensed background galaxies. Credits: X-ray: NASA/CXC/CfA/M. Markevitch et al.; Lensing Map: NASA/STScI/ESO WFI; Magellan/U. Arizona/D. Clowe et al./Optical image: NASA/STScI

This is explained because the hot gas does have some mutual braking effects, while the dark matter just carries on moving after the collision and separates from the rest. The term "bullet" refers to the shape of the baryonic mass distribution in the smaller of the two clusters. This separation of baryonic matter and dark matter can be well explained on the assumption that the gas in the two clusters acts to slow down the motion of the baryons by friction, while the dark matter interacts only by gravity and speeds through the encounter. This is one of the most convincing observations showing that dark matter is "real" and not just an effect of some kind of distortion of gravity. To be fair the proponents of the MOND, modified Newtonian gravity theory as an alternative to dark matter have been able to give a qualitative account of the Bullet cluster, but the numbers in the dark matter scenario are easier to fit.

5.3.7   Spectr-RG Maps the Whole Sky in X-Rays

Technical progress is an intrinsic part of a subject as intrinsically technical as X-ray astronomy. In 2019 a joint project between Russian and German astrophysics institutes, the Spektr-RG satellite, was launched. The satellite contains two complementary sets of instruments, eROSITA a mapper produced by the German group and ART-XC a detector of point sources at high energy from the Russian group. The high performance of eROSITA comes from a set of seven parallel grazing incidence telescopes, which provide a very wide field, 0.9° on a side, in soft X-rays between 0.3 and 7 keV, but quite good angular resolution: 18 arcseconds, across the full sky. ART-XC also has 7 grazing incidence telescopes, but as it deals with higher energies (6–30 keV) it has a field of only just over 0.5° on a side, and a resolution of 45 arcseconds.

Most of the structures seen in Fig. 5.23b belong to our Galaxy, and include well known sources we have met, or will meet several times in this book: the Cygnus superbubble, Cassiopeia A, the Crab nebula, the Orion Nebula, but we can also see clusters of external galaxies: the Coma Cluster, the Virgo Cluster, the Fornax cluster, the Perseus Cluster, which are seen as single extended sources, because their intergalactic media are emitting thermal X-rays at million degree temperatures. We can even see a much larger but more distant structure, the Shapley supercluster of galaxy clusters.

a

b

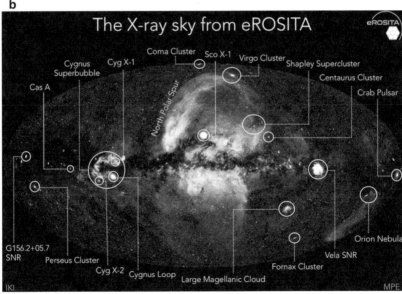

Fig. 5.23 (a) The 7 telescopes which form the eROSITA instrument on board the Russian-German satellite Spektr-RG, launched in 2019. Each is a Wolter-type grazing incidence instrument. This configuration allows the observation of a wide field at high angular resolution. Credits: MPE/IKI/Roscosmos. (b) The whole sky mapped by the

Further Reading

Riccardo Giacconi: The Nobel Lecture. "The Dawn of X-ray Astronomy". https://
www.nobelprize.org/uploads/2018/06/giacconi-lecture.pdf

Riccardo Giacconi , History of X-ray telescopes and astronomy. Experimental
Astronomy, Vol. 25, p 143. 2009 On Springer Link

Keith Arnaud "An Introduction to X-ray Astronomy". Power point presentation:
https://heasarc.gsfc.nasa.gov/docs/xrayschool-2007/arnaud_intro.pdf

Aneta Siemiginowska "An Extraordinary View of the Universe : The Use of X-ray
Vision in Space Science". Métode Science Studies Journal, Vol. 7 p. 163. 1917.
web link: https://ojs.uv.es/index.php/Metode/article/view/8819/9770

Fig. 5.23 (continued) Russion-German X-ray telescope eROSITA. The map is a projection in coordinates based on our Galaxy, with the plane of the Milky Way across the centre. The false colours represent lower energy X-rays (0.3–0.6 keV) in red, intermediate energies (0.6–1.0 keV) in green, and the higher energies (1–2.3 keV) in blue. The image has been smoothed to avoid contrasts with the zones still less well mapped. Credits: Max Planck Institut für Extraterrestrische Physik: Jeremy Sander, Hermann Brunner and the eSASS team), IKI (Russia): Evgeny Churazov, Marat Gilfanov

6

Gamma Ray Astronomy from Space

6.1   Why Observe the Universe in Gamma-Rays?

Gamma-rays are the most energetic part of the full electromagnetic spectrum. Modern physics is very used to the quantum mechanical concept which describes electromagnetic radiation in terms of both waves and particles (photons). As the energy of the radiation goes to higher values, the wavelengths get shorter and each of the photons in the radiation gets more energetic. Astronomical sources produce gamma-rays with energies much higher than anything we can find on Earth. In 2019 a gamma ray photon with the highest energy ever detected, coming from the Crab Nebula, was measured at 450 Tera-electron volts (TeV) which is 4.5×10^{14} electron volts (eV) To give an idea of what this means, the rest mass of a proton when converted to energy (using $E = mc^2$) is only around 900 eV, so these high energy cosmically produced gamma rays carry a powerful punch. We are fortunate that the Earth's atmosphere is sufficient a shield to protect us from direct incidence by the high energy gamma rays, which would certainly cause damage to our cells if they arrived *en masse*, and would probably have prevented the emergence of life. This implies, though, that if we want to study the sources of gamma-rays to find out about the production mechanism but also to explore the general physical conditions of these sources throughout the universe we need to use satellite observatories. This is generally true, but as we can see in the chapter on cosmic rays there is an alternative technique for studying high energy gamma-rays based on the showers of sub-atomic particles they produce when they hit the upper atmosphere. The present chapter will concentrate on the satellite observations, which deal with the relatively lower energy range of

© Springer Nature Switzerland AG 2021
J. E. Beckman, *Multimessenger Astronomy*, Astronomers' Universe,
https://doi.org/10.1007/978-3-030-68372-6_6

gamma-rays below some 300 GeV, a limit set by the limited size of the tele-scopes which can be flown on satellites.

6.2   What, Specifically, Can We Hope to Learn from Gamma-Ray Astronomy

There is a general rule, which we see throughout this book, that the higher the temperature of a body or system, the shorter the wavelengths of the radiation which it emits. But gamma-ray production does not rely on this rule, because it would mean that gamma-rays would be produced simply by the hottest bodies. This "thermal" production does not work in practice for them. The hottest widely distributed bodies which do emit a little gamma radiation are white dwarfs, but at surface temperatures of over 100,000 K they emit prin-cipally in X-rays. Gamma-rays are produced in quantity in more "exotic" pro-cesses which allow the concentration of large amounts of energy in single photons. The key science questions for which gamma-ray astronomy can give us interesting answers are (1) The origin of cosmic rays (CR). The mecha-nisms which accelerate particles to the highest energies can be probed using the gamma-rays they also emit. They travel towards us directly along straight paths, which is a big advantage in identifying sources. The questions asked are "What are these sources"? and "What mechanisms are accelerating the cosmic rays"? To resolve these problems we need to combine gamma-ray and CR observations. (2) Relativistic flows. The CR's are often accelerated in relativis-tic outflows, such as winds or jets. The types of sources are active galactic nuclei, (AGN), gamma-ray bursters (GRB) and compact binary stars in the Galaxy, which are commonly found as X-ray sources. Collimated jets are often formed by the accretion of matter onto the most compact objects, (neutron stars and black holes), and gamma-rays are a good way of exploring these rela-tivistic jets. (3) Cosmological questions. The gamma-rays from AGN can be used to put constraints on intergalactic magnetic fields and fields of electro-magnetic radiation on the largest scales, to probe voids, and in tests of quan-tum gravity. (4) Dark matter searches. One hypothetical way to look for dark matter is that the products of the self-annihilation of two colliding dark mat-ter particles could be gamma-rays, and theorists have derived possible spectral energy distributions for them. The idea is that they should concentrate towards zones of the deepest gravitational potential, which implies the nuclear zones of galaxies. Among the candidates are weakly interacting massive particles, "WIMPS" with masses in the range 0.01–10 TeV. If they exist, the gamma-ray signatures of their annihilation could allow us to find them in cosmic sites, and to connect their properties with direct measurements in laboratories such as CERN here on Earth. Another dark matter candidate, the axion, could be revealed by the distortion of the gamma-ray energy spectrum of distant sources by the interaction of the axions with the gamma-rays in magnetic fields en route to the observer.

6.3   A Brief Time-Line of Gamma Ray Observations from Space

The earliest gamma-ray observations were made with NASA satellites. The first gamma-ray telescope in orbit was on the Explorer 11 satellite in 1961. It detected less than 100 cosmic gamma ray photons. These came from all directions, and suggested the existence of a uniform gamma ray background radiation, which could be caused by the interaction of cosmic rays with the interstellar gas. In 1967 a gamma-ray detector on the OSO3 satellite picked up gamma-rays from within the Galaxy and also from outside, and major steps forward were taken with SAS-2 which flew in 1972, and with ESA's COS-B (1975–1982) (Fig. 6.1).

These two satellites confirmed the presence of a gamma-ray background, and produced a map of the sky, detecting a number of point sources where the sites of gamma-rays were concentrated. However their instruments did not have good angular resolution and could not identify these sources with stars or with any particular type of cosmic systems. One of the most spectacular discoveries in gamma ray astronomy came in the late 1960s and early 1970s, when a set of satellites from the US Department of Defense, the Vela gamma-ray satellites, designed to detect flashes of gamma rays from nuclear bomb tests, detected flashes coming from outer space. These are the now famous gamma ray bursts (GRB's) which can last from a fraction of a second to a few minutes. These were studied from a number of satellite and space probes, including the Soviet Venera probe and the Pioneer Venus Orbiter. Their sources were unknown and some theorists speculated that they might be a halo of neutron stars around the Galaxy, as they came from all directions. However in 1996 a combined effort by the ESA BeppoSax gamma-ray detector and the Hubble Space telescope showed that at least one of the sources was in an external galaxy. In 2004 NASA launched the Swift satellite, whose aim was to locate GRB's very quickly and report their position on the sky, so that

Fig. 6.1 ESA's COS-B gamma-ray satellite on test in the laboratory. Credit: ESA

they could be followed up at other wavelengths. By 2010 Swift had reported 500 GRB's and the number detected has kept on increasing with time.

The two major gamma-ray observatories by NASA have been the Compton Gamma-Ray Observatory, CGRO, launched in 1991 and de-orbited in 2000 because of a failure in a stabilising gyroscope, and the Fermi Gamma-Ray Space Telescope, launched in 2008 and still in operation. Both of these satellites gave considerable information about the most energetic activity in the universe including solar flares, neutron stars, supernovae and black holes. Another satellite with gamma-ray capability is AGILE, launched by the Italian Space Agency in 2007 and containing both gamma ray and X-ray detectors.

6.4 Gamma-Ray Telescopes

The most difficult aspect of building a gamma-ray telescope is that gamma-rays penetrate all solid surfaces so that mirrors for gamma-rays cannot be made. This means that the normal optical methods which can be used in all wavelength ranges from the longest radio waves to the shortest X-rays cannot be applied to building a gamma-ray telescope. There are three physical processes which can be used, either singly or together, for detecting gamma-rays in a telescope system. They are the photoelectric effect, Compton scattering, and pair production. In the photoelectric effect the incoming gamma ray removes an electron from an atom in the detector, and this electron can be directly detected by appropriate circuitry. The effect is the same for gamma rays as it is for much longer wavelengths, such as visible light. This effect is only applicable to the lower energy range of the gamma ray spectrum, for energies less than 1 MeV, and is not used in practice for astronomical measurements. In Compton scattering a gamma-ray interacts in a material with a free electron, (which is not bound to any of the atoms in the material), and loses energy, which is transferred to the electron, and can be detected and measured. This type of interactions is most efficient in the middle range of gamma-ray energies. In the pair production process, the high energy gamma-ray converts to two particles, an electron and a positron, and the positron later interacts with an electron in the detecting material, in such a way that the two particles annihilate, giving rise to new gamma-rays, which can themselves Compton scatter, or produce a photoelectric effect. Either way the energy of the original gamma-ray can be estimated. The basic detector for gamma-rays in the range from 100 keV to 10 MeV is a "scintillation counter" which is made of material in which light is produced by charged secondary particles produced either by the photoelectric effect or by Compton scattering, plus a photodetector in which the light is converted to an electrical signal. A much used scintillator material is sodium iodide, activated with thallium, NaI(T). A gamma-ray telescope must not only detect the gamma-rays, but also reject the scintillations produced by the charged particles in the constant flux of cosmic rays. This is done by surrounding the gamma-ray detector with a vessel made of another type of scintillator sensitive to cosmic rays.

In a Compton telescope there are two layers of detectors separated by a distance of a couple of metres. The upper detector Compton scatters a primary gamma-ray photon; its energy is inferred by measuring the recoil energy of the electron which scattered it. The scattered ray is then detected by absorption in another Compton scatterer in the lower level. A diagram of this is shown in Fig. 6.2.

The upper detector is thin so the gamma rays are scattered once, and the lower detector is thick so that the scattered ray is fully absorbed. The detectors are surrounded by shields to protect them from cosmic rays. Assuming that the scattering process in the upper detector deviates the gamma-ray by only a small angle, the information obtained from the two detectors is enough to estimate the energy of the incoming gamma-ray photon, and also its direction on the sky within a narrow ring, which for the Comptel detector on the CGRO was some 4° in diameter. By making a number of observations with

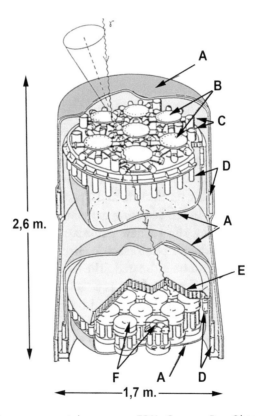

Fig. 6.2 Comptel gamma-ray telescope on ESA's Gamma Ray Observatory. The schematic shows the two sets of scintillators described in the text. Credit: NASA, Public domain, via Wikimedia Commons

different pointings on the sky the position of the source can be determined using the intersection points of the rings. A Compton telescope of this type works best in the gamma-ray energy range from 10 to 30 MeV. For the higher range of 30 MeV–300 GeV a pair-production telescope is used. The central component of this is a spark chamber, a device invented and used in particle physics. When a gamma-ray hits the upper part of the chamber and interacts to produce an electron-positron pair, the particles fly on through a set of parallel plates in a vessel filled with a suitable gas, usually a neon-ethane mixture. This particle triggers a high voltage pulse across the chamber, and this causes the particle to leave a narrow track of ionised gas as it passes through. This track defines the direction of flight and gives the desired information about the position of the source on the sky. The trigger detection system effectively defines a field of view on the sky, and the energy of the gamma-ray is measured using a thick block of NaI(T) scintillator at the bottom of the telescope. The gamma-ray which reaches this block has lost a significant fraction of its initial energy in the pair production and annihilation process, and in producing the ionisation of the track, but the relation between the energy detected at the bottom of the telescope and the energy of the incoming gamma-ray is calibrated from tests prior to flight. The whole telescope is surrounded by a vessel of thin scintillator, sensitive to cosmic rays but transparent to gamma-rays, whose action is to stop the cosmic rays from reaching the spark chamber.

A prime example of a spark-chamber gamma-ray telescope was EGRET, which flew together with the Compton telescope on NASA's CGRO which was operational on orbit form 1991 to 2000, with a complement of four different gamma-ray telescopes. In Fig. 6.3 we can see the CGRO being placed into orbit.

The most recent major gamma-ray satellite telescope is NASA's Fermi, launched in 2008, and still operational. It has two main instruments. The first is a large array telescope, (LAT) working on similar principles to EGRET but with a collecting area and sensitivity which make it 30 times more sensitive, and with an order of magnitude superior resolution for localising sources on the sky. It also covers an much wider energy range, from 20 keV to 300 GeV, compared with an upper limit of 30 GeV for EGRET. The second instrument is a gamma ray burst monitor (GMB), with improved performance compared with a similar instrument, BATSE on the CGRO. Fermi has made a large number of varied types of observations over more than a decade, some of which we will discuss below when we consider the scientific objectives of gamma-ray astronomy.

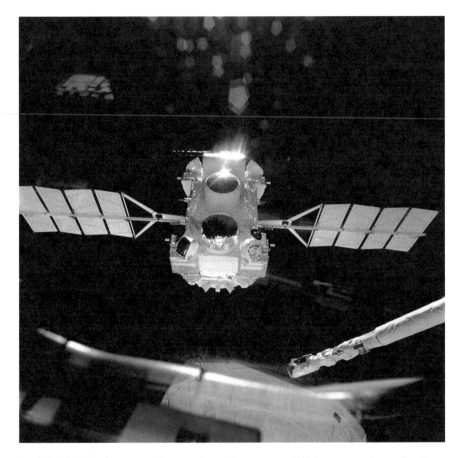

Fig. 6.3 NASA's Compton Gamma-Ray Observatory drifting away from the Space Shuttle while being placed on orbit, June 4th 2000. Weighing 17 tonnes it was the heaviest astrophysical payload to have been launched at that time. Credit: STS-37 crew/ Compton Science Support Center/NASA

6.5   What Gamma-Ray Astronomy Is Telling Us

6.5.1   The Gamma-Ray Sky

Figure 6.4 is a general picture of the gamma-ray sky at GeV energies, published in 2015. It shows the three main components observable in gamma-rays: (1) the diffuse emission from our Galaxy, generated by the interaction between the general flux of local cosmic rays (CR) and the interstellar gas, and radiation fields; this is concentrated towards the centre plane of the Galaxy, (2) Individual sources of gamma-rays, and (3) a faint glow of diffuse isotropic emission, observed at all Galactic latitudes. The proportions of these three components varies as the energy band changes. At GeV energies four fifths of

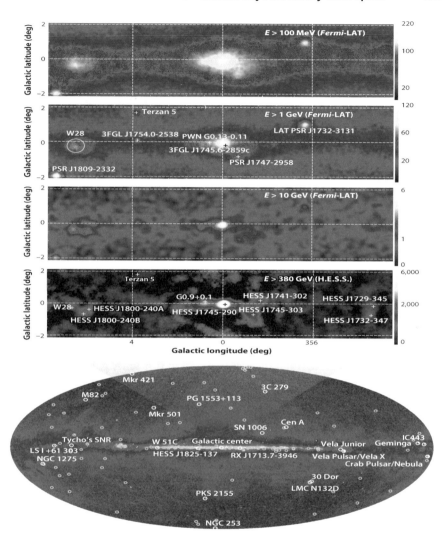

Fig. 6.4 The gamma-ray sky at GeV energies as mapped by the Fermi-LAT imager on the Fermi satellite. Bottom elliptical panel. The high energy gamma-ray sky as seen by the Fermi-LAT satellite instrument after 6 years of observing. The colours are coded for intensity per pixel for gamma-rays above 1 GeV energy. The positions of TeV sources detected with ground-based gamma-ray telescopesare marked with cyan circles, or white circles for a selected set of prominent gamma-ray sources. The colour scale is logarithmic with a range of three orders of magnitude between dark blue and white. Top three rectangular panels: Zoomed-in view into the Galactic Centre region, corresponding to a size of 3500 × 900 light years separated into separate plots each with a different minimum energy; these energies are marked on the maps. On the middle panel several bright sources have been marked, showing a variety of source types. Credit: Funk, S. 2015, Annual Reviews of Nuclear and Particle Science, Vol. 65, pp 245–277

all the photons are from the Galactic diffuse emission, at TeV energies the individual sources dominate, because their spectra contain more high energy photons. Even so the Galactic diffuse emission has been detected at TeV energies from ground based gamma-ray telescopes.

As the angular resolution of these telescopes is not very high, we do not yet know whether this comes from cosmic ray production or from as yet unresolved point sources. Although there are a number of different point source types, dominated by different mechanisms of gamma-ray production, they all have in common spectra whose power falls off quite steeply to higher energies, so for a given telescope sensitivity we expect to find more low energy sources than high energy sources. In the 4th Fermi-LAT catalogue of sources, published in March 2020 there are over 5000 sources observed in the range 40 MeV–1 TeV to a high degree of significance (4σ). Of these over 3000 have been identified as active galaxies of the blazar class, i.e. highly variable at optical and/or radio wavelengths, 239 are pulsars. Seventy five of the sources have been shown to be spatially extended with detailed modelling, and over 350 are also identified as extended. However 1300 of these gamma-ray sources have not yet been identified with any counterparts at other wavelengths. It is clear that the limitations on resolution make it more difficult to identify sources, and to separate individual sources. So in this sense gamma-ray astronomy is less complicated than astronomy at other electromagnetic wavelengths, and in this sense it is somewhat comparable to radioastronomy in the 1960s and 1970s.

6.5.2   Gamma-Rays Tracing the Origin of Cosmic Rays: Supernova Remnants

All gamma-ray sources are also potential sources of cosmic rays (CR's), and in particular of CR protons, but unlike protons or electrons they are not deviated significantly in their paths to the observer by magnetic fields. Of course they may interact with particles en route, and be deviated considerably, just as light photons can interact with cosmic dust, for example, but in spite of this if a significant beam of gamma-rays reaches us it will point us at a source of cosmic ray particles. The flux of gamma-rays produced by protons and their related particles, the hadrons, is determined by the CR density and the density of the gas with which the CR's collide to produce the gamma-rays within the source. The flux of gamma-rays produced by electrons is determined by the electron density and the radiation fields in the source. To distinguish the two types of sources we normally need to combine gamma-ray observations with radio and X-ray observations in order to pin down the needed parameters. We know that, in general terms, to produce photons and particles with the huge energies observed the source must be a violent one. This leads us directly to supernova remnants as the suspected sources. We know from radio and optical observations that supernova remnants (SNR's) contain very high energy electrons and magnetic fields. But the electron densities are not high enough to produce the observed gamma-ray fluxes, so they must be produced by protons. This fits with the high magnetic fields because the best way we know of to produce these fields is in a shock front under the pressure of accelerated protons. So the gamma-rays have been an essential element in helping us to understand the shocks in supernova remnants. Young SNR's with large measured magnetic fields are the best targets to search for gamma-ray emission produced by protons and sources of cosmic rays with energies of up to 10^{15} eV, not yet in the uppermost range observed, but still very high. SNR's close to a dense interstellar cloud, a molecular cloud for example, where the expanding shock wave drives into the cloud, are expected to produce a high flux of hadronic gamma-rays.

These scenarios where gamma-rays should be produced have been confirmed by the observations. Apart from pulsars, which are known to produce CR's due to their high energy electrons, SNR's are the commonest gamma-ray sources within the Galaxy. Several tens of these sources have been well observed. The largest group of those producing gamma-rays in the GeV range are those interacting with molecular clouds, and the second group is that of young SNR's, which are less luminous in this range, but more luminous in the higher energy TeV range.

Figure 6.5 shows a set of gamma-ray spectra from Fermi-LAT. These spectra are just the intensity of the gamma-rays as a function of energy. They have been coded in colours to show the different types of sources. The most "famous" sources are the very young SNR's. For example Tycho's supernova, which was seen in the sky in 1572 and has been the subject of considerable recent research, has a spectrum extending into the higher energy range, and Cassiopeia A, or Cass A which was one of the first sources to be discovered by the infant science of radioastronomy, in 1948, is estimated to have produced visible light peaking in principle some 300 years ago, although this light was almost certainly hidden by a deep dust cloud because there are no records of people observing a supernova in the expected period. The three principal categories of SNR's in the gamma-ray range are well illustrated by the spectra.

Figure 6.6 shows maps of one of the intermediate aged SNR's, comparing a map in X-rays obtained by the XMM-Newton X-ray satellite, a map in the GeV energy range obtained with Fermi-LAT, and a map in the high, TeV

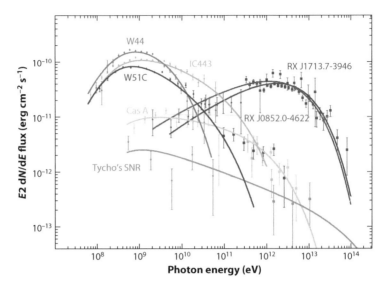

Fig. 6.5 Spectra of some of the brightest gamma-ray sources in the Galaxy. Measurements mostly from the Fermi-LAT telescope, of the fluxes of high energy cosmic rays from supernova remnants (SNR's) The source names are the traditional object names taken from measurements at other wavelengths, radio, optical and X-ray. The known young (less than 1000 years old) SNR's are in green. They show lower fluxes but these extend over the GeV and TeV ranges. Somewhat older SNR's (~2000 years old) shown in red, have their energy peaked in the TeV range. Older SNR's (~20,000 years) interacting with molecular clouds are shown in blue. These have strong fluxes, but at lower energies, peaking in the GeV range. The curves are fits to the observed data points using theoretical models in which the gamma-rays are produced by cosmic ray protons interacting with the surrounding medium. Credit: Funk, S. 2015, Annual Reviews of Nuclear and Particle Science, Vol. 65, pp 245–277

energy range from the HESS ground based telescope. In fact the different spectra of the sources shown in Fig. 6.5 are all consistent with the general scenario of SNR's being the prime sources of high energy cosmic rays, but the experts in the field do not have an easy time deciding between the detailed production mechanisms for the gamma-rays, for example whether they are produced by protons or electrons in any given case. This is because in the zone of any SNR there is interstellar material of multiple phases. In a type II SNa the massive parent star has formed quickly, has blown off a massive wind, and has then exploded. The gas immediately around the star is atomic and ionised even before the explosion, but it is probable that the molecular cloud which gave birth to the star or the star cluster has not been dissipated and is still there a bit further out to receive the violent impact of the expanding SN Shell. The Shell itself can contain a strong magnetic field, so there is scope for all the possible mechanisms for gamma-ray production. Future

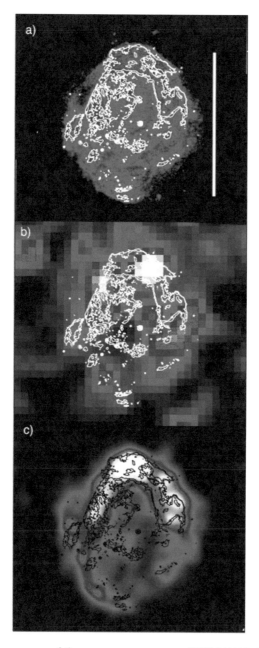

Fig. 6.6 High energy maps of the supernova remnant (SNR) RX J1713.7-3946 (a) upper map in X-rays from the XMM-Newton satellite observatory, (b) middle map in GeV gamma-rays from the Fermi-LAT satellite telescope, and lower map (c) in TeV gamma-rays taken at the HESS ground-based observatory using the air shower technique (see Chap. 9 for more details). Credit: Funk, S, 2015, Annual Reviews of Nuclear and Particle Science, Vol. 65, pp. 245–277

observatories both in space and on the ground will undoubtedly resolve some of these complexities.

6.5.3 Gamma-Rays from Pulsar Winds

Pulsar wind nebulae (PWNe) are the most numerous single class of TeV gamma-ray sources in the Galaxy. It is of particular interest that in this type of sources the TeV gamma-ray luminosity can be strong, yet the GeV and the X-ray luminosities can be weak or absent. When Fermi-LAT was used to search systematically for GeV emission from the neighbourhood of discovered TeV sources, only 20% of them had measurable GeV fluxes. The high energy PWNe are understood to be related to young and energetic pulsars which power magnetised nebulae with fields between a few microgauss and a few hundred microgauss. But the TeVe luminosities do not show a correlation with the mechanical power of the pulsar, which can be measured from its deceleration rate the "spin-down" power. The youngest pulsars show a good match between the gamma-ray morphology and that observed in X-rays with the Chandra or XMM-Newton satellites, while for older pulsars, (remember, a dynamical age can be deduced by measuring the pulse rate and also the rate at which it is slowing down so that the time interval between pulses is growing) the TeV source, from the wind directly surrounding the pulsar, is displaced from the centre of the extended X-ray emission. This is either because the pulsar is moving with respect to the initial expanding SNa shell which produces the X-rays (and also can produce GeV gamma-rays), or because the shell is expanding into an inhomogeneous surrounding medium and so loses its original symmetry with respect to the pulsar. One of the current mysteries of TeV gamma-ray sources in the Galaxy is why almost one third of them are lacking in radio wave or X-ray counterparts. A possible solution to this could be provided by the older PWNe. The electrons in the expanding cloud around the pulsar could yield high energy gamma-rays in the TeV range by scattering the ambient photons in the general background radiation, either the cosmic background or the diffuse infrared background due to dust, or even diffuse visible starlight. This scattering process is referred to as Inverse Compton scattering (see the description of Compton scattering earlier in this chapter). But the electrons coming from the pulsar lose their energy rapidly in the magnetic field of the star itself, and cannot emit sufficient X-rays to be detected. To delve into the physics of these sources it will be necessary to obtain higher resolution TeV images of them and this is one of the programmes to be solved by a new generation of ground-based gamma-ray telescopes, embodied in the CTA, the Cherenkov Telescope Array, which we will see in Chap. 9.

6.5.4 Gamma-Ray Detection from Other Galaxies: Escape of Cosmic Rays into the Intergalactic Medium

The question of whether the GeV gamma-rays in other galaxies are produced by similar mechanisms to those which operate in our Galaxy is an active field for current study. Gamma-rays have been detected from many nearby galaxies, above all those with active star formation. This is reasonable considering the mechanisms for production mentioned in the previous section. We detect current star formation in other galaxies by the presence of clusters of massive, luminous blue stars which must be young, as the more massive a star the shorter is its lifetime. By the same token these clusters will, in general, contain SNe and SNR's because these massive stars end their short lives as type II SNe. A significant part of the energy of a SNa is emitted in cosmic rays and gamma-rays. In fact the modelling of these processes suggests that the time scale for this energy escape is quite short, so that during a burst of star formation there should be a rough equilibrium between the gain of energy in the CR's due to input from the supernovae and loss of energy by gamma-radiation and the escape of the CR's away from their formation sites. A study by Fermi-LAT of over 60 nearby galaxies selected for their abundant molecular gas, and therefore likely sites for distributed star formation, shows a clear monotonic relation between gamma-ray luminosity and both infrared luminosity and radio continuum luminosity.

All of these parameters are basically determined by the star formation rates in the galaxies, more star formation more cosmic rays. Figure 6.7 shows the relation between the integrated energy observed in the GeV range and observed measurements of the luminosity in the radio range at 1.4 GHz frequency, i.e. the 21 cm wavelength continuum. These relationships may be used, in the long term, by finding the most detectable of any of these multimessenger parameters, as standard candles for evaluating the large scale structure of the universe. Although there is much more observing and interpreting to be done, a general conclusion from this is that most of the energy in CR's in most galaxies eventually escapes into the intergalactic medium, (IGM) where their paths are determined by the magnetic field. In this way the IGM is a reservoir for high energy CR's, which have accumulated over gigayears and have very slow decay rates because their low densities lead to very long mean free paths for collision or interaction.

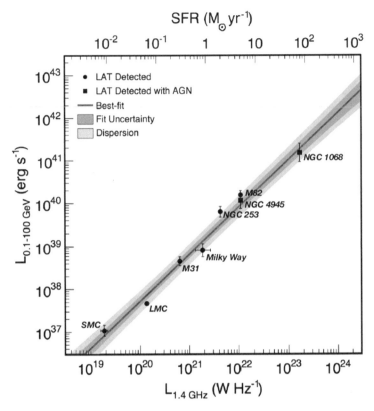

Fig. 6.7 GeV Gamma-Ray Emission from Normal and Starburst Galaxies plotted against the emission at 1.4 GHz radio emission from the same sources. Some of the data in the original diagram have been removed for visual simplicity. Credit: Knödlseder, J., 2013, in Cosmic Rays in Star- Forming environments, Astrophysics and Space Science Proceedings, vol. 34, 169–191/IAC-UC2

6.5.5 The Fermi Bubbles

One of the first major tasks that the Fermi satellite set out to accomplish was a to produce a gamma-ray map of the whole sky, and its large field of view made this task relatively straightforward. The Galaxy dominates this map, and after integrating the measurements for the first 18 months in orbit the Fermi-LAT telescope obtained a spectacular result. From the centre of the Galaxy, in opposing directions vertically above and below the plane two huge elongated bubble structures were observed. In order to separate them from the diffuse emission caused by the CR's from the sources in the galactic plane interacting with the off-plane gas the gamma-ray specialists had to use measured maps of the atomic hydrogen and the molecular hydrogen obtained at radio wavelengths and to model the quantity of gamma-rays produced by the interaction of these CR's with the gas. They calculated a "diffuse gamma-ray map" of the Galaxy, which was primarily meant to help in valid detection of point sources, but turned out to be very good for picking out the bubbles, which showed up well with the same shapes in each of the separate energy bins used within the GeV range. The bubbles extend to over 30,000 light years above and below the plane. The most probable scenario is that they are the result of a burst of accretion onto the supermassive black hole at the Galactic centre, either of a massive gas cloud, or of a star cluster whose stars were torn apart by the immense gravitational field of the black hole.

Figure 6.8 is a cleaned TeV image of the Fermi Bubbles made from the Fermi-LAT observations. These bubbles were examined in a most interesting observational experiment using the Hubble Space telescope in which the UV spectrometer COS was used to observe distant quasars behind the bubbles and obtain the velocity fields and the chemical composition of the gas throughout the bubbles using its absorption spectrum.

6.5.6   Gamma Ray Bursts

In 1967 there was already a gamma-ray satellite in orbit. It was the Vela satellite, not designed for astronomical observations, but for the military purpose

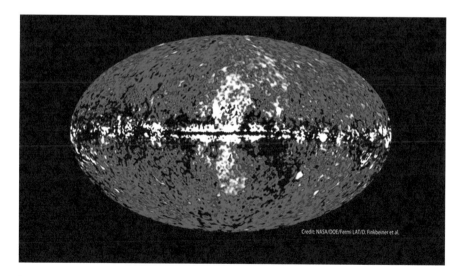

Fig. 6.8   Map produced by the Fermi-LAT satellite telescope showing the Fermi Bubbles integrated over the GeV range of gamma-ray frequencies colour coded for intensity. We see the concentration of gamma-rays in the Galactic plane, and compact sources distributed around the sky. The two bubbles extend respectively over 30,000 light years above and below the plane, and are probably due to a burst of energy emitted from just above the event horizon of the supermassive black hole when matter fell into it. Credit: NAS/DOE/Fermi LAT/D. Finkbeiner et al

of detecting the test explosions of nuclear weapons, so it was aimed at detecting very short period pulses of gamma-rays. To the surprise of the military, it detected very rapid bursts which clearly originated from outside the Earth. As the existence of Vela was not revealed to the general public, it was not until it was declassified in 1973 that astronomers learned of these gamma-ray bursts. There followed a steady stream of detections and, naturally, of theoretical models and scenarios to explain the bursts, which included collisions between neutron stars, and even collisions between comets. From 1991 the Compton Gamma Ray Observatory, (CGRO) and its Burst and Transient Source Explorer (BATSE) showed that the distribution of the bursts on the sky is isotropic, for which the most plausible explanation is that the sources are from outside the Galaxy, as sources within the Galaxy would be concentrated in its plane (Fig. 6.9).

In 1997 the Italian-Dutch satellite BeppoSax detected a gamma-ray burst, and its X-ray camera detected an afterglow in X-rays, which allowed a position on the sky to be determined. Follow- up on the ground using the optical

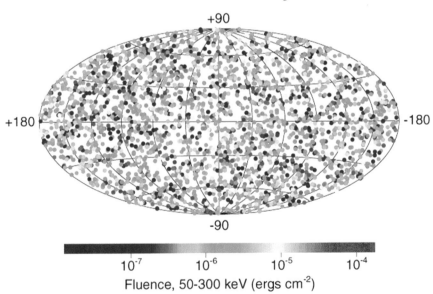

2704 BATSE Gamma-Ray Bursts

Fluence, 50-300 keV (ergs cm^{-2})

Fig. 6.9 Distribution on the sky of the gamma-ray bursts detected by the BATSE detector on NASA's Compton Gamma Ray Observatory (CGRO) during 19 years of operation. The uniform distribution on the sky strongly suggests extragalactic origins for the bursts. This has been confirmed by multi-wavelength observations (see text). Credit NASA /CGRO Science Support Center

William Herschel Telescope on La Palma (Canaries) found a fading optical counterpart 20 h after the initial gamma-ray burst, and showed a faint distant galaxy in that direction, although too faint to measure its recession velocity, i.e. a value for its distance. In the following months, BeppoSax detected another burst, GRB 970508 within 4 h of its discovery, which allowed optical astronomers to obtain a spectrum, and determine its redshift: $z = 0.835$. This clearly placed it in a galaxy at 6000 million light years from us. A year later the gamma-ray burst GRB 980425 was followed within a day by a bright supernova, SN 1998bw, in the same location, giving a strong clue about the nature of the bursts, and their sources. The general interest in these bursts led to a number of specialised satellite missions to detect them. The CGRO ceased operations in the year 2000, and BeppoSax stopped functioning in 2002, but two special missions, HETE-2 (2000–2006) and notably Swift (2006–) detected many gamma-ray bursts. Swift has a very sensitive gamma-ray detector for the bursts, as well as X-ray and optical cameras for the afterglows, which can be slewed swiftly to observe the sources. The Fermi mission, which we have seen in a more general context, has a gamma-ray burst monitor, which detects several hundred GRB's per year, and the Fermi-LAT instrument detects the brightest of them at high energies. Astronomers networked at observatories on the ground are now ready to use robotic telescopes, triggered by messages from gamma-ray satellites, to make follow-up observations within minutes, even seconds, of the initial discovery.

Short period gamma-ray bursts are defined as those with duration less than 20 s, and make up around one third of all the bursts. Until 2005 follow-up had not been quick enough to detect afterglows, leaving their nature a mystery. After that date an increasing number of them has been located on the sky, but often towards objects with very little or no star formation, such as elliptical galaxies and central zones of clusters of galaxies. This points away from massive stars, and shows that these events are different from the longer bursts. Theorists suggested, from the short timescale, that they might be the result of the merger of two neutron stars or of a neutron star with a black hole. Such events would yield a very rapid gamma-ray burst, and possible longer duration afterglows, later, when the particles released impinged on fragments of the disrupted stars. These conjectures were dramatically confirmed by the follow-up of the gravitational wave source GW170817, the first source of gravitational waves whose timescale could be attributed to a neutron star merger rather than the merger of two black holes, which had been the source of all the previous detections. This detection was followed only 1.7 s later by a short gamma-ray burst. We will meet this event again in the chapter on gravitational wave astronomy.

There have been far more observations of long gamma-ray bursts, as these are much easier for multi-wavelength follow up. Virtually all of them are in galaxies with high star formation rates, and are attributable to the supernovae caused when a massive star collapses and is blown apart after exhausting its nuclear fuel. Some of these last as long as a few hours, and there is speculation that these may be due to some entirely different process, such as the complete disruption of a massive star by the tidal effect of a supermassive black hole at the centre of a galaxy.

All the types of events which produce gamma-ray bursts are among the most highly energetic events known in astronomy. Their apparent luminosities are high, but many of them are also at great distances, which means that their absolute luminosities must be extremely high. We calculate the luminosity of any object by measuring the amount of energy falling on a square metre here on Earth, and by knowing the distance to the object we can derive the total luminosity, the power it must emit to project the energy per second per square metre which we measure. If we do this for some of the most energetic gamma-ray bursts we find extraordinarily high values, equivalent to the conversion of the mass of a star into energy during the burst. This is almost impossible to produce by known physical processes. So we must assume that the gamma-rays are not emitted isotropically, but beamed into a relatively small solid angle and that we must lie within that beam. Beamed processes are frequent in astrophysics, occurring where high energy jets of particles are confined by high magnetic fields.

6.5.7   Dark Matter Particle Annihilation Searches

One of the most intriguing possibilities for gamma-ray observations is the prediction that the particles which make up the dark matter which comprises, according to the currently preferred scenario, some 80% of the mass within our Galaxy, and some 20% of the mass of the universe, may decay by mutual interaction, yielding electromagnetic radiation. According to models which attempt to quantify this the probability of the mutual interactions should vary according to the square of the dark matter density, and this density should be highest in the centres of galaxies. However the degree of concentration of both baryonic ("normal") matter and dark matter in the centre of a galaxy is hard to determine. One of the main criticisms of the current cosmological models has been that the original models predicted a very sharp cusp in the central density of galaxies, which should be detectable from the gravity of the matter distribution using the shape of the rotation curve as we approach the centre, and the predicted shape was not observed. Various "fixes" have been introduced into the original cosmological galaxy formation simulations to remove the cusp. The simplest fix is that the simulations used dark matter alone, as this dominates the total matter content and makes for simpler modelling, but that in the centre the baryons should dominate so the cusp models were inadequate. Another fix is that intense star formation in the central zone gives rise to winds and supernovae which sweep out much of the baryon content, dragging some of the dark matter with it. One consequence of any of these scenarios is that the dark matter concentration at the centre is less than the original models had predicted, so that the rate of self-annihilation and the production of gamma-rays is reduced. Nevertheless the centre of our Galaxy is still considered a good place to look for a signal from dark matter annihilation. There is a prominent source of gamma-rays at the Galactic Centre which coincides with the centre of rotation and the strong compact radio source Sagittarius A*, where the 4 million solar mass black hole is situated. This source was detected with the HESS ground-based gamma-ray telescope in the TeV range, and confirmed by observations with other ground-based telescopes, notably the MAGIC telescope, so that a spectrum in TeV gamma-rays could be plotted. Later measurements into the GeV range were made with Fermi-LAT. The shape of the spectrum is not that predicted by any of the models invoking dark matter particle annihilation as the source of the gamma rays. In fact there is not yet a consensus on the nature of the source, which could be either a baryon or a leptón source associated with the supermassive black hole, or even a pulsar wind. Observations at higher angular resolution to pin down the exact position of the source, and relate it to sources at other wavelengths are needed, and might be provided by the new large telescope array for detecting high energy gamma rays from astronomical sources, the CTA. Searches for gamma-rays coming from dark matter annihilation at the centres of dwarf galaxies in our Local Group of Galaxies have so far shown similar inconclusive results to those from the Galactic centre. Clearly compared to astronomy in the lower energy ranges gamma-ray astronomy is still in quite an early development phase, and will undoubtedly be a field to watch in the future. High energy gamma rays are intimately linked with the production of cosmic rays, and the techniques of high energy gamma ray observations from the ground have much to do with the physics of the sources producing cosmic rays, so I decided to include ground-based gamma-ray astronomy in the chapter on cosmic rays and we will meet it in Chap. 9.

Further Reading

"Twelve Years of Education and Public Outreach with the Fermi Gamma-ray Space Telescope" Lynn Kominsky et al. in 4th Fermi Symposium, Monterey, California. 2012. Downloadable from the preprint server astro-ph with reference https://arxiv.org/abs/1303.0042v1

"Fourteen Years of Education and Public Outreach for the Swift Gamma-Ray burst Explorer Mission" Lynn Kominsky et al. in eConf. 2014. Downloadable from astro-ph with reference https://arxiv.org/abs/1405.2104v1

A NASA website mainly for young people: https://imagine.gsfc.nasa.gov/science/toolbox/gamma_ray_astronomy1.html

7

Neutrino Astronomy

7.1   What Is a Neutrino?

The neutrino is a particle which has almost no mass, and almost no interactions with any other particles. Nevertheless, its existence is necessary for physics to work at all, and its role in the universe has become a source of valuable information for astronomy. It is important because it plays a key role in one of the four basic interactions, or forces, in physics. These four are the electromagnetic force which regulates most of the processes in our lives: atomic and molecular processes, as well as light and all the types of radiation; the gravitational force, the major attraction between massive bodies including stars and planets; and the two forces which act within the atomic nucleus, the strong force and the weak force. When radioactivity was discovered and investigated, the nuclei which suffered spontaneous radioactive decay were found to emit three principal types of objects: gamma rays, which are "very high energy light" (electromagnetic radiation), alpha particles and beta particles. The alpha particles were later found to be helium nuclei, containing two protons and two neutrons, while the beta particles were shown to be electrons. The force which effectively acts between the neutrons and the protons within a nucleus is called the strong force, and this binds stable nuclei together. Within a nucleus both protons and neutrons are stable, but a free neutron outside a nucleus decays spontaneously into a proton and an electron after about ten minutes, in a process which is controlled by the fourth force, the weak force. This process puzzled the physicists in the 1920s when it was first found, because it appeared to behave in an impossible way. The total energy of the two particles produced was always less than the energy of the original

© Springer Nature Switzerland AG 2021
J. E. Beckman, *Multimessenger Astronomy*, Astronomers' Universe,
https://doi.org/10.1007/978-3-030-68372-6_7

neutron, and also the sum of the momenta of the two particles was less than that of the original neutron, even though the sum of their masses was equal to the original mass. This seemed to break two physical laws, the conservation of energy and of momentum. The situation was not so different from our present puzzle about what is the missing "dark" matter which forms most of the mass of the galaxies. In 1930 the theoretical physicist Wolfgang Pauli suggested that the only way to resolve the paradox would be the existence of a particle with no mass, or very little mass, which was able to carry away energy and momentum from the decay. It would not only have to be of very low mass, but it could hardly interact with anything once it had escaped, because these particles had not been detected. Pauli named this elusive particle the neutrino, (neutral but smaller than the neutron) (Fig. 7.1).

For over 25 years the neutrino remained a purely hypothetical particle, but in 1956 Clyde Cowan and Frederick Reines (Fig. 7.2) prepared an experiment in which they bombarded water with a flux of 10^{17} neutrinos per second per square metre that were being emitted from a nuclear reactor at the Savannah River plant in South Carolina (Fig. 7.3).

Fig. 7.1 The theoretical physicist Wolfgang Pauli who predicted the existence of the neutrino Credit: W. Dieckvoss, Hamburg. Reused with permission from the Pauli Archive, CERN (PAULI-ARCHIVE-PHO-231-1)

Fig. 7.2 Frederick Reines, experimental physicist who, with Clyde Cowan first demonstrated the existence of the neutrino. Courtesy: University of California Regents

To help ensure that they were not detecting other reactions induced by cosmic rays, they placed their detector tanks 12 m below the ground and 11 m from the reactor. Each interacting neutrino was predicted to react with a proton (hydrogen nucleus) in the water creating a neutron and a positron (positive electron). When a positron approaches an electron they react quickly, destroying themselves to produce a gamma ray which can be detected by a nearby scintillator, producing a flash of detectable light. To confirm that this was due to a neutrino Cowan and Reines (Fig. 7.3) also placed a tank of cadmium chloride solution in their experiment. When the cadmium is hit by the neutron produced by the neutrino it emits a gamma ray which is also detected by the scintillator, five microseconds later than the first gamma ray. This arrangement provided a unique signal for detecting neutrinos, and an average rate of three detected neutrinos per hour was found over a period a several months. To clinch the observation, the experimenters closed down the reactor, and the detection rate went down effectively to zero. The success of this experiment produced the Nobel Prize for Reines in 1995 (unfortunately Cowan had died in 1974 and Nobel prizes are not awarded posthumously) (Fig. 7.2).

There are three types of neutrinos known, and they are incorporated into the wider current scheme of particle physics. They are the electron neutrino, the type discovered by Cowan and Reines, the muon neutrino, first found by Lederman, Schwarz and Steinberger in 1962 and the tauon neutrino first detected by a large collaboration of researchers at the Fermilab in the year 2000. All three types of neutrinos have two peculiar properties not shared by other fundamental particles. Firstly they are their own antiparticles (we need not go further into this as it would need a long diversion into particle physics,

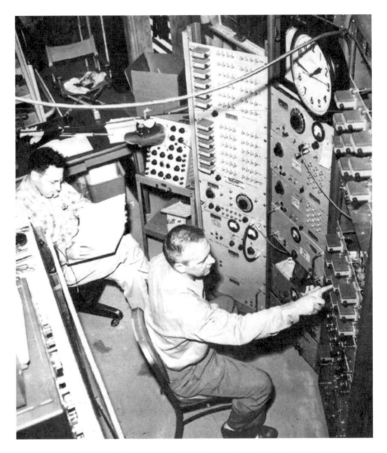

Fig. 7.3 Reines and Cowan in their laboratory 12 m below the reactor at the Savannah River Power Plant, where the first detection of neutrinos was made on August 27th 1956. Courtesy: CERN archives

not appropriate in this book) and secondly they can transform spontaneously, passing from one type to another. This latter phenomenon is known as neutrino oscillation, and it plays an interesting role in the subject of solar neutrino physics as we will see shortly. The probability of an oscillation is higher when the neutrino passes through matter than when it is crossing the near vacuum of interstellar space.

The fact that neutrinos interact so weakly with other particles is both a problem and an advantage. It is a problem for the obvious reason that they are difficult to detect. One way of expressing this is in terms of their effective cross-section. This is a concept often used in molecular, atomic, and particle physics to show how likely it is that a given particle interacts with others. We can think of it in terms of a circular target, such as a dart- board or a shooting target, which we are trying to hit. The bigger the target the bigger the chance of hitting it. The cross section of a particle is the effective area it presents when we are shooting another particle at it, and also when we are shooting it at another particle. The cross-section of a neutrino is very small indeed, around 10^{-48} square metres. Another way of looking at this is to calculate how far we could fire a beam of neutrinos into a block of lead before they are stopped by interactions with the lead nuclei. The result is that only half of them would be stopped within a light year. A consequence is that to detect neutrinos, we need either a very intense flow of them, or a very large detector, or a very long time, or preferably all three. But the advantage of this very weak interaction is that if neutrinos are produced in the hottest, densest, or most active places in the universe, they can escape from their site of origin and cross space to our detectors with the minimum of interference en route. We know that the stars are fuelled by nuclear fusion reactions, mostly the fusion of hydrogen to helium, and that two neutrinos are emitted for each helium nucleus produced, (this is for the most common reaction; we will see more about nuclear reactions in the Sun in the next section). The neutrinos travel straight out from the centre of the Sun without further interactions and so if we can capture even the tiniest fraction of them here on Earth we will have direct information about the Sun's centre. We are unable to study the centre of the Sun directly using electromagnetic radiation, all the light we detect comes from a thin layer at the surface. This penetrating property of neutrinos is gradually being brought into service as an effective astronomical tool to study the objects of much higher energy than even the centres of normal stars.

7.2 Solar Neutrinos

7.2.1 The Total Output of Solar Neutrinos

It is not difficult to estimate the output rate of neutrinos from the fusion reactions in the centre of the Sun. By measuring the solar constant, how much solar energy per second is falling on the Earth (corrected for absorption by the atmosphere) and knowing the mean distance of the Earth from the Sun we find the total power output of the Sun is almost 4×10^{26} watts. We also know the energy made available when four hydrogen nuclei (protons) effectively combine to produce a helium nucleus (using the most famous formula in physics $E = mc^2$ and the small loss of mass of the helium nucleus compared to the four protons) is close to 4×10^{-12} joules, and that one watt is one joule per second. This gives us an estimated rate of 10^{38} fusions per second, If two neutrinos are emitted per fusion the rate of neutrino emission from the Sun is 2×10^{38} neutrinos per second. This number is very big, and even when we take the distance of the Sun from the Earth into account, the number of neutrinos per second which are passing constantly through a given area here on Earth is a large number: some 10^{15} neutrinos per square metre per second. It is interesting to make an estimate of the effect of this flow of neutrinos on the human body. Taking the cross sectional area of a standard human body seen from above as a fifth of a square metre, and the weight of water in this standard body as 50 kg the expected interaction rate of the solar neutrino flux will be around one per year. Our bodies are not very practical detectors of solar neutrinos, and perhaps more important the damage they do to us is utterly negligible (compared for example to cosmic ray impacts).

7.2.2 Some Details About How Solar Neutrinos are Produced

There are a number of reactions within the Sun which produce its energy, and give neutrinos as by-products. From the point of view of solar astronomers trying to detect them the most important differences between these reactions are the different ranges of energies of the neutrinos produced. This is because different methods of detection are sensitive to different ranges. These differences played an important role in the early discoveries of solar neutrinos leading to a set of puzzles whose solution has led to deeper understanding of both the Sun and the physics of the neutrino.

Figure 7.4 shows the numbers of neutrinos produced within the Sun which arrive at the Earth per second. Each curve shows the product of one of the chains of nuclear processes which are occurring in the Sun's core, giving rise to its total energy output which we see finally as light from the photosphere, the "solar surface". By far the most frequent of these processes is the p-p, the proton-proton chain reaction, which produces some 85% of the Sun's energy. As we can see from the figure these neutrinos have the lowest range of energies, so detectors which detect only high energy neutrinos will not detect them.

7.2.3 Detection of Solar Neutrinos

The experiment which was the first to detect neutrinos from the Sun was a project headed by the experimentalist Raymond Davis Jr. ("Ray Davis") and theorist John N. Bahcall. They designed their experiment in order to use a

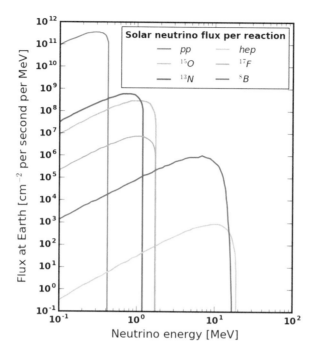

Fig. 7.4 The flux of solar neutrinos at the Earth produced by the different types of nuclear reactions in the core of the Sun, as functions of the energy of the neutrinos released in each type of reaction. Credit: CC BY-SA 4.0 <https://creativecommons.org/licenses/by-sa/4.0>, via Wikimedia Commons (using data from Bahcall et al. Astrophysical Journal Letters, Vol 621 p L85, 2005)

reaction in which a neutrino interacts with a chlorine atom, producing radioactive argon and an electron. The chlorine presents a relatively high cross-section for the reaction. The detector was a 380 cubic metre tank of the dry-cleaning fluid perchloroethylene, chosen because it allows a high concentration of chlorine nuclei, and because it allows the gaseous argon produced to be separated quite easily. This was done by bubbling helium through the liquid every few weeks to collect the atoms of argon, and as the isotope of argon produced is radioactive, it decays back to chlorine and emits X-rays which allow the number of the radioactive atoms to be counted. The experiment was performed in the Homestake Gold Mine in South Dakota, (Fig. 7.5) almost 1500 metres below the ground in order to shield the detectors from cosmic ray particles and gamma rays which are constantly impinging on the Earth, and whose signals would have drowned out the faint neutrino signal. As we know, operating at this depth offers virtually zero barrier to the Neutrinos (Fig. 7.5).

Fig. 7.5 Construction of the percholoroethylene neutrino detector tank in the Homestake mine in 1966. Ray Davis supervising the work. By permission: Brookhaven National Laboratory

The experiment was the most practical initial way to look for solar neutrinos, but it had the disadvantage that the reaction of the chlorine nuclei requires a neutrino of energy more than 0.8 MeV so that, as you can check in Fig. 7.4, only a small fraction of the neutrino flux could be detected. But such a large overall flux did make the experiment a success (Davis was awarded the Nobel Prize for this in 2002, although by that time Bahcall was no longer alive and could not share it). The theorists, using a model of the nuclear reactions within the Sun, were able to compute how many of the detectable neutrinos (those with energies higher than 0.8 MeV) should be detected in a given time, and this prediction was a little more than one neutrino per day. As time went by, and the number of detections accumulated steadily, the numbers of neutrinos were consistently about one third of the numbers predicted by Bahcall's models. The initial opinion of most researchers in the field was that either Davis' detections, or Bahcall's models were in error. But further experiments were performed to extend possibilities of detecting solar neutrinos and they all gave similar results, which did not depend on the energies of the neutrinos detected. These experiments were performed in different countries and used a variety of detecting methods, extending the energy range of observable neutrinos. One of the experiments was GALLEX, carried out in the Gran Sasso Laboratory in Italy, some 3000 metres below the ground, by an international team including French, German, Italian, Israeli, Polish, and American scientists. It used a solution containing over 30 tonnes of the element gallium, which reacts with a neutrino to form a radioactive isotope of the element germanium. The germanium decays in a few days, and its decay products can be easily detected. This method has the virtue that it detects neutrinos with energies above 0.4 MeV, so that it can detect the neutrinos produced by the p-p reaction in the sun, by far the dominant energy producing reaction, with by far the largest production of neutrinos. The experiment ran for 6 years between 1991 and 1997 and made an average of 0.75 neutrino detections per day. So the results were firmly based on reasonably large numbers, and still gave a value of only one third of the numbers predicted by the best solar models. This discrepancy became known at the solar neutrino mystery and it puzzled physicists and astronomers for over three decades.

7.2.4   The Solar Neutrino Mystery

The solution to the mystery came from the results of two different neutrino observation experiments: Superkamiokande in Japan, and SNO in Canada. We will be meeting Superkamiokande in the next section because its work was not directly concerned with solar neutrinos, so here we will look only at SNO. The Sudbury Neutrino Observatory, SNO was sited in the Vale Creighton Mine at Sudbury, Ontario. It was designed to detect solar neutrinos in a tank containing 1000 tonnes of heavy water. Heavy water is water where the normal hydrogen, H, whose nucleus contains just one proton and nothing else, is substituted by deuterium, D, whose nucleus contains a proton and a neutron bound together. The formula for heavy water is D_2O, like that for water H_2O. The detector was surrounded with normal water to give it some protection from radiation which could be confused with the neutrinos. The solar neutrino detectors prior to SNO had all been sensitive to only one of the three types or "flavours" of neutrinos, the electron neutrinos, but SNO could detect all three types: electron, muon, and tau neutrinos which all react with the deuterium nuclei, although in different ways. In addition the SNO detector covered a wider energy range, making the statistics of its detections more reliable. When its first results were published in 2001 it was clear that it was detecting at the full rate predicted by the solar models, and not at only one third of the rate, as had been found in the previous studies. The answer to the solar neutrino mystery was then at hand. The models predicted that all the neutrinos emitted by the sun should be electron neutrinos, so how could SNO have also detected muon and tau neutrinos? A quantum mechanical effect, "neutrino oscillations" had in fact been surmised as a possible answer. Due to this effect the neutrinos in a general flow change type from one to the other of the three types, varying with time on quite short timescales, and reaching an equilibrium where equal numbers of all three types are present in the flow. So in the time taken for the solar neutrinos to reach the Earth two thirds of the original electron neutrinos had mutated into either muon or tau neutrinos, and could not be detected by the Homestake detector or by GALLEX or by a number of other experiments prior to 2001. This phenomenon of neutrino oscillations implies that a neutrino is different from a photon, in that it has a finite mass, something not consistent with the previous standard model in elementary particle physics, but which had been predicted as a theoretical possibility as early as 1957. Astronomical observations show that this mass must be very small indeed, less than one millionth of the mass of an electron. The importance of the discovery of neutrino oscillations for the foundations of physics was indicated by the award of the 2015 Nobel Prize for Physics to Art McDonald, the director of SNO, which he shared with Takaaki Kajita, the director of Superkamiokande, which we will now go on to describe (Fig. 7.6).

7.2.5 Kamiokande

In the early 1980s a group of Japanese physicists set up an experiment to try to measure the lifetime of the proton. In some theories of particle physics the proton is not perfectly stable, and could decay into other particles, but with a very long lifetime, greater than the age of the universe. The physical reasoning behind the experiment was straightforward; if the average time for a single proton to decay is very long, in order to have a reasonable chance to detect a measurable number of decays one has to assemble a large number of protons and surround them with detectors which can detect the decay products, and then wait. The experiment consisted in placing a large tank of water deep underground, and surrounding it with photomultiplier detectors. If a proton in the water decays, it will produce a relativistic particle, a particle which will travel faster than the speed of light in water, and this will generate light, due to an effect known as the Cherenkov effect, which the photomultipliers can detect. We will encounter this effect in Chap. 9 when we talk about

Fig. 7.6 The neutrino detector in the SNOLAB, in the Sudbury mine in Ontario, where the full spectrum of solar neutrinos was detected, leading to the discovery of the phenomenon of neutrino oscillations, and showing that the neutrino must have a finite rest mass (unlike the photon) albeit less than one millionth of the mass of the electron. This shows the construction of the Photomultiplier support structure which surrounded the acrylic vessel for the original SNO experiment. This has since been updated to the currently active SNO+ experiment. Credit: SNOLAB

cosmic ray and gamma ray astronomy. The tank and its detectors were sited deep underground in a mine, in a district called Kamioka, and the experiment was named Kamioka NDE with the initial NDE standing for Nucleon Decay Experiment. Placing it underground gave it shielding from spurious detections due to cosmic rays at ground level. The cylindrical water tank contained 3000 tons of pure water, and there were 1000 photomultiplier tubes each of 50 cm diameter, attached to its inner surface. It was first constructed between 1982 and 1983, and although it did not detect any proton decays it did set the best lower limiting value to the lifetime of the proton, which has been progressively increased with its upgraded successor Superkamiokande, and now stands at 1.6×10^{34} years, (compared with the age of the universe, a mere 13.8×10^9 years).

After the first disappointment at not detecting proton decay with Kamkiokande the physicists at the University of Tokyo who had designed it decided that it could be used as the basis for a new experiment to detect and characterise solar neutrinos. To do this they had to make a number of changes, purifying the water to reduce the amount of the radioactive element radon, and building an outer jacket of water to shield the inner water chamber from radiation from the surrounding rock and from the remanent cosmic rays which reach the depths of the mine. The upgraded system, Kamiokande-II operated from 1985 with its mission of detecting neutrinos from the sun. The water detector gave an advantage that in this type of detectors the reaction used to detect neutrinos is their interaction with electrons. This is a scattering process analogous dynamically to the collision between billiard balls, and the direction of the scattered electron's path responds to the direction of the incoming neutrino, and the energy it receives is also a measure of the energy of the neutrino. It took a couple of years to reduce false detections from local radon to an acceptable minimum. Then systematic data on neutrino detections could be gathered, and after a year and a half the experimenters could claim that they were detecting neutrinos from the direction of the sun, and could measure their rate of arrival. This turned out to be less than one half of the predicted rate, in fair agreement with the Homestake results, and at that time still under the shadow of the solar neutrino mystery. By the 1990s physicists were beginning to wonder whether this problem might be due to neutrino oscillations, and the Japanese physicists, in collaboration with colleagues from a number of partner countries, decided to build a detector which might resolve the question, based essentially on the Kamiokande design.

7.2.6   Superkamiokande

This was a monster version of the earlier detector, with 50,000 tons of pure water surrounded by some 11,200 photomultiplier tubes. Figure 7.7 gives an idea of the scale of the interior, before the water tank was introduced.

An outer detector shielded the main tank from cosmic ray muon particles, and could also determine their incoming tracks. The observations of the solar neutrinos over the full energy production range with Superkamiokande was one of the key steps in the resolution of the solar neutrino problem, showing that the neutrinos oscillated among the three flavours. It also studied the high energy range of neutrinos produced in the earth's atmosphere by the impacts of very high energy cosmic rays, and showed that these also are subject to the phenomenon of oscillations.

In Fig. 7.8 we can see an "image" of the Sun produced by the Superkamiokande detector. Firstly we should know that the colours in the image are, of course, false colours used to represent the intensity of the source, which is brightest at the centre and fades radially outwards. Secondly we must

Fig. 7.7 A general view of Superkamiokande before the water tank neutrino detector was introduced. Credit; Kamioka Observatory, ICRR (Institute for Cosmic Ray Research), The University of Tokyo

note that the size of the image is some eighty times the size of the normal visible image of the Sun, because the distribution of the detections on the sky is very strongly "blurred", defocused by the weak angular response of the system. There are no "neutrino lenses" or "neutrino mirrors" technically possible. If there were a way to focus the neutrinos, yielding a sharp image, we would see a bright source covering only the central 10% of the optical disc, corresponding to the projection of the hot core where the nuclear reactions produce the Sun's energy, and its neutrinos.

7.3 Neutrinos from Supernova 1987a

We have seen that Kamiokande was an experiment which "failed" by not detecting proton decay, and that the imaginative use of the facility allowed physicists to detect solar neutrinos. But a different achievement of Kamiokande was even more striking because unexpected. On February 23rd 1987 the Canadian astronomer Ian Shelton and the Chilean Oscar Duhalde found a new bright object on a photographic plate of one of our nearest neighbour galaxies, the Large Magellanic Cloud (LMC) in a programme to search for

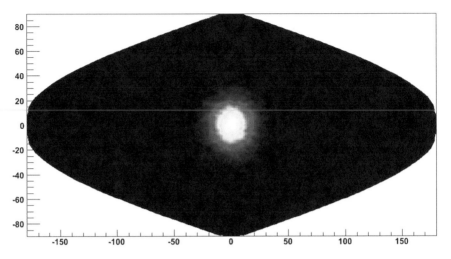

Fig. 7.8 "Image" of the sun in neutrinos taken with the Superkamiokande neutrino telescope, during 5200 days of integration between 1996 and 2016. Scale in degrees. The large angular size is because the detection method cannot produce good angular resolution on the sky The image diameter is some 40 degrees which is the effective angular response of the instrument. (Half-width of the response at 10 MeV is about 20 degrees). The colours represent neutrino counts, with respect to the position of the centre of the Sun, white high values, yellow intermediate values red low values, black background level. The Sun's visible disc would be just a point at the centre of the image. Credit; Kamioka Observatory, ICRR (Institute for Cosmic Ray Research), The University of Tokyo

variable stars, at the Las Campanas observatory in Chile. It was bright enough to be seen by the naked eye (magnitude 3–4), and was reported 4 h later by the amateur astronomer Albert Jones in New Zealand. Its brightness and known distance identified it as a supernova, the first to be seen with the naked eye since Kepler's famous supernova in 1604. Three days later John Bahcall, whom we mentioned as one of the leaders in the Homestake solar neutrino experiment, and colleagues, sent a letter to the journal Nature suggesting that Kamiokande-II ought to have been able to detect up to 50 neutrinos of the 50,000, 000,000 per square cm predicted to pass through the earth from this supernova. The Kamiokande team looked back into their data and found a burst of 12 neutrinos arriving on February 23rd at 7 h 35 m and 35 seconds UT (Universal Time) within a time range of 13 seconds. This was 3 h before the first optical record of the supernova at 10:37:55 UT. The detection was confirmed by two other neutrino detector experiments, the IMB detector (named for three collaborating physics groups (Irvine-Michigan- Brookhaven) underground near Lake Erie in the US, which detected 8 neutrinos, and the Baksan Neutrino Observatory (BNO), a Russian experiment in the Caucasus

Mountains, detected 5. A graph showing the times of arrival and the energies of the detected neutrinos at all three sites is shown in Fig. 7.9a, and a graph showing the detections at Superkamiokande alone in Fig. 7.9b. The latter plot gives the background level of neutrino detections due to local cosmic ray interactions, and shows very clearly the detection of the neutrinos from the supernova.

It is interesting to note that the Large Magellanic Cloud is in the southern sky, not visible from almost all of the northern hemisphere, while the three observatories, Kamiokande-II, IMB, and BNO, are all in the northern hemisphere. This highlights the very weak interaction of the neutrino, as all the detected neutrinos passed through a large part of the Earth before reaching their detectors in the opposite hemisphere. Although the detection of 25 particles may not seem to be very important, the neutrinos from SN 1987a marked the beginning of a new era, the era of neutrino astronomy. Masatoshi Koshiba, the director of the Kamiokande-II experiments was awarded the Nobel Prize in 2002 for the discovery, a prize which he shared with Ray Davis, for the first detection of solar neutrinos at the Homestake Mine. It seems clear that John Bahcall, who was the theorist behind the Homestake experiments and also directly triggered the Kamiokande team to search for their neutrino signal, would have shared the prize had he still been alive in 2002. (The 2002 prize had a third recipient, Riccardo Giacconi, one of the pioneers in X-ray astronomy).

What were the results for astronomy and physics of the SN 1987a neutrino detections? The first conclusion was about the physics of supernovae. The neutrinos allow us to make inferences about processes within the supernova which cannot be directly accessed by electromagnetic radiation. The theoretical models for this type of supernova predict that 95% of the energy released is in the form of neutrinos, and give an estimate for the number of neutrinos produced in the explosion, which should take place over the amazingly short period of a few seconds. We have a reasonable estimate of the distance of the Magellanic Cloud, close to 165,000 light years, so it is possible to estimate how many neutrinos should pass through, and then how many should be detected in a tank of water at Earth, with the size of Kamiokande-II. This is in fact what Bahcall and his colleagues did when they sent their estimate for publication. The fact that the resulting detection was in the ball-park of their estimate gives reasonable support to the theoretical models.

The second conclusion was an estimate of an upper limit to the mass of the neutrino. The Kamiokande measurements gave the energies and the arrival times of the neutrinos in the burst. Particles which travel at the speed of light (photons) have no rest mass but physicists know, because of the observed

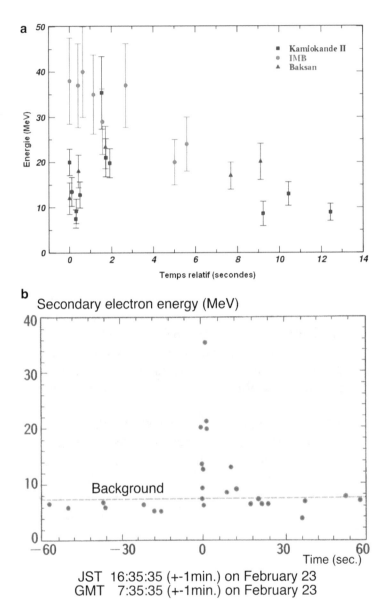

Fig. 7.9 (a) A plot of the arrival times and energies of the neutrino burst detected at three different sites from Supernova 1987a. Although the energies have significant error bars, all the detections are significant, and collectively the result is unquestionable. (b) Plot of the detection of the neutrinos from SN 1987a at the Superkamiokande detector, showing the background level of neutrino counts, and indicating the clarity of the detection Credit; Kamioka Observatory, ICRR (Institute for Cosmic Ray Research), The University of Tokyo

phenomenon of neutrino oscillations, that neutrinos must have a rest mass, a feature common to particles in general, such as electrons, protons, and neutrons. This must be very small, and implies that neutrinos released from a supernova explosion travel at speeds very close to the speed of light. By comparing the arrival times and energies of the neutrinos, and using information from the optical observations about the differences in emission time the astronomers showed that a neutrino cannot have a mass bigger than one ten thousandth of the mass of an electron. The time between the release of the neutrinos in a supernova explosion and the release of the light from the optical surface of the explosion is a matter of seconds (the neutrinos are released first). This marks an interesting contrast with the Sun, and other stable stars. A neutrino produced in a nuclear fusion reaction in the sun's core takes some 2.3 seconds to reach the surface, travelling at very close to the speed of light, and a little over 8 minutes to travel to the Earth. A gamma ray photon, produced in the same reaction in the sun's core immediately interacts with nearby nuclei. It is absorbed and re-emitted as it makes its way to the surface, losing energy in keeping the rest of the sun at its equilibrium temperature. When it reaches the photosphere, the emitting surface, its wavelength has been reduced to that of visible light, appropriate to the photospheric temperature of 5500 K. The time taken from its emission in the core, to its release from the photosphere can be computed from models at some 100,000 years, although we can see that it is not "the same photon" as that produced in the centre. The photon then takes just over 8 min to reach the Earth. In its passage from solar core to solar surface direct information about the physical conditions in the core is lost, so neutrino research will give us useful insights, as the numbers grow. The same is true of supernova explosions, though since 1987a no other supernovae have occurred near enough for us to observe their neutrinos.

By far the major part of the information gathered about SN 1987a has been obtained by conventional methods: optical, radio, and X-ray astronomy. We know from its light curve, how the brightness varied with time, that it was a Type-II supernova, caused when a star with a mass in a range above 8 times the mass of the sun has exhausted the nuclear fuel in its centre, and suffers a dramatic collapse, followed by the explosion. Theorists and observers together have worked on models for this type of supernovae, which are observed nowadays in many distant galaxies. After the light from the explosion was seen, astronomers looked on images of the area in the LMC where it had occurred, and located the star which had been its progenitor. This turned out to be a blue supergiant, Sanduleak -69202a which surprised the astronomical community, as Type-II supernovae were supposed to have red supergiants as their predecessors. This point is elaborated in the chapter on ultraviolet astronomy.

7.3.1   SNEWS, the Supernova Early Warning System

Core collapse supernovae are those produced when a single massive star, of some 8 times the mass of the Sun or more, has burned through its fuel cycle and the reactions in the core start to absorb energy rather than produce it. This produces a dramatic collapse under gravity, leading to an explosive disintegration of the star. These supernovae produce not only enormous bursts of electromagnetic radiation, but also a huge flux of neutrinos. The neutrinos escape directly from the core of the star travelling very close to the speed of light without hindrance from the body of the star, while the photons take several hours to emerge from the body of the star. This means that the neutrinos will arrive here significantly earlier than the photons as was the case for SNa 1987a. But core collapse supernovae also emit neutrinos at higher energies so the detection techniques augment those used for SNa 1987a. The SNEWS (Supernova Early Warning System) project is a collaboration between seven neutrino experiments around the world each capable of detecting these higher energy neutrinos. Among them is Superkamiokande, in Japan, and the Ice Cube detector in Antarctica, which we will describe later in this chapter, as well as further detectors in Italy, China, and another Japanese experiment. The SNOlab mine in Canada is also incorporated, by way of the HALO detector, comprising a set of ^{3}He neutron detectors embedded in 79 tonnes of lead (the HALO acronym is from Helium and Lead Observatory). The original SNO detector will also be used within the SNEWS project. The protocol of SNEWS is that if two or more of the experiments detect a burst of high energy neutrinos, an alert is sent to the whole astronomical community to await electromagnetic radiation from the supernova a few hours later. The HALO detector is sensitive to electron neutrinos, while the other detectors are sensitive to antineutrinos, so additional information of physical interest should be forthcoming. However, patience has to be the watchword, as no more than two or three core collapse supernovae per century are expected to occur within the Galaxy, and the number of neutrinos reaching us from extragalactic sources in general is too small for reliable detection.

7.4   Neutrinos, Antineutrinos and the Asymmetry of the Universe

While I was writing this chapter Kamiokande participated in another fundamental experiment with neutrinos, which is well worth the mention here. Physicists used the J-PARC accelerator at Tokai on the east coast of Japan to create beams of neutrinos and fire them in the direction of Superkamiokande, nearly 300 km away. The distance is sufficient for the phenomenon of neutrino oscillations to show up. Even though the accelerator laboratory produces billions of neutrinos per second in the beam, the detector registers very few indeed. During a decade of runs only 90 electron neutrinos and 15 electron antineutrinos have been detected. The initial beams are of muon neutrinos and muon antineutrinos, and the results indicate that muon neutrinos are more likely to oscillate to electron neutrinos than muon antineutrinos to electron antineutrinos during the flight time of the experiment. This asymmetry of behaviour has not yet been established by the experiment at a level totally incompatible with pure chance, but the result is highly indicative. If it is confirmed by upgraded experiments between the same accelerator and detector this difference in behaviour between neutrinos and antineutrinos, technically termed CP-violation (violation of charge-parity symmetry) may hold the key to the asymmetry of the universe which has caused matter to dominate over antimatter and explain why not all the energy in the Big Bang was turned into gamma-rays. In short this result could be a key to the existence of the baryonic universe, with its atoms, stars, and galaxies.

7.5   High Energy Neutrino Astronomy

We have seen how Kamiokande made use of water as a neutrino detector, and how in general neutrino telescopes have to be profoundly shielded to minimise the possibility of confusion by cosmic ray particles. In addition, the detection of solar neutrinos, which are of relatively low energy, requires the water to be purified so that the effect of the radioactive gas radon does not produce spurious detections. But for high energy neutrinos, predicted to be produced in violent events such as supernovae, and the mergers of white dwarfs, neutron stars, and black holes, the latter precaution is not necessary. This led physicists to the idea that water in sufficient quantity could be both a detector and a shield, a principle which was used in Superkamiokande. The natural extension of this principle would be to put a neutrino telescope into very deep water, and this has been carried out in two ways. Firstly several neutrino detectors have been installed at the bottom of the sea, notably in the Mediterranean, where countries active in physics and astronomy, France, Italy, Spain, and Greece, have coastlines (Fig. 7.10).

These projects include ANTARES, off the French coast near Toulon, which was completed in 2008 and comprises twelve strings of detectors in a volume of water roughly 200 by 200 metres based at 2.5 kilometres below the sea

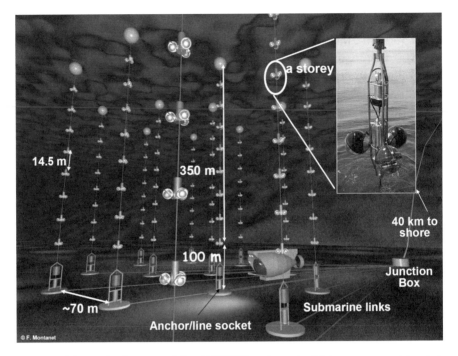

Fig. 7.10 Schematic diagram of the ANTARES underwater high energy neutrino telescope Courtesy of F. Montanet/CNRS/IN2P3/ ANTARES COLLABORATION

surface. It was considered a prototype and it showed its paces by detecting tens of neutrinos per day. All of these were produced in the Earth's atmosphere by incident cosmic rays, but neutrinos originating in the Sun or from cosmic sources much further away were not detected. The NEMO and the NESTOR neutrino telescopes were built in the Mediterranean by Italian and Greek based consortia respectively during the same period. A major collaborative project KM3NeT is in preparation by a large consortium of European countries together with Morocco and Australia, comprising a 2 km cubic array of neutrino detectors based on the Cherenkov principle. The Russians have built their underwater neutrino detector BDUNT at a depth of 1.1 km within lake Baikal starting operations in 2003 and upgrading with additional detectors progressively since then, with the aim of completing a cubic km volume detector in 2020. All of these detectors are designed similarly, as a set of strings each string being of length a hundred metres to a kilometre long, depending on the specific project. On each string are placed at intervals a set of neutrino detectors which in fact detect the light emitted by particles, essentially muons, travelling faster than light in the water around the strings. These particles are produced when a neutrino of sufficient energy impacts a proton in the water. We will take a detailed look at these detectors and how they work as we examine the most successful neutrino telescope, built within the Antarctic ice.

The most powerful high energy neutrino observatory to date is operating deep in the Antarctic ice. It is aptly named Ice Cube and comprises 86 vertical strings each with 60 detector modules, suspended between 1450 and 2450 m below the surface of the ice, in a virtually cubic configuration (Fig. 7.11).

The holes for the strings were made using hot water drilling. Each detector module, (DOM) has complex electronics whose aim is to record the flash of Cherenkov light emitted when a neutrino interacts with a hydrogen nucleus in a water molecule. Depending on the type of neutrino, the interaction gives rise to an electron, a muon, or a tau. If the energy of the incoming neutrino is high enough, the particle travels faster than the speed of light in ice, and this produces the Cherenkov radiation, which is recorded in the surrounding DOMs which digitise the information and send it to the surface along their cables. The muons have the greatest penetrating power, and the Cherenkov radiation they produce can be picked up by a number of the DOMS, and can be used to reconstruct the track of each muon through Ice Cube, in the framework of the timing of each detection, which is accurate to two nanoseconds. As the track of the muon is quite well aligned with the track of the incoming high energy neutrino, the information can be used to point to a position on the sky of an astronomical source. We should remember that the sources detected in this way are in the northern sky! As well as using the water (frozen, but this is of no consequence) as a detector and a partial shield, the Earth itself is a principal element in Ice Cube. Neutrinos are continually being produced in the atmosphere by the impact of cosmic rays and Ice cube detects far more of these than it does astronomically produced neutrinos. But the vast majority of them can be rejected because they are travelling downwards. Even so the small fraction coming from cosmic ray impacts on the atmosphere at the antipodes of Ice Cube do dominate the number of detections from the northern hemisphere. One way of distinguishing neutrinos from specific astronomical sources is by their energy, those of the highest energy are expected to come from sources such as active galactic nuclei. Another way is their directionality. Atmospheric neutrinos, produced by cosmic rays of moderate to fairly high energy come towards the detector from all directions, whereas the neutrinos of highest energy have greater directionality. One interesting observation about the cosmic ray produced neutrinos is that Ice Cube detects the shadow of the Moon. There is a slight, but statistically significant reduction in the neutrino arrival rate from positions on the sky occupied by the Moon, which is blocking a fraction of the incoming omnidirectional cosmic rays. Figure 7.12

a

b

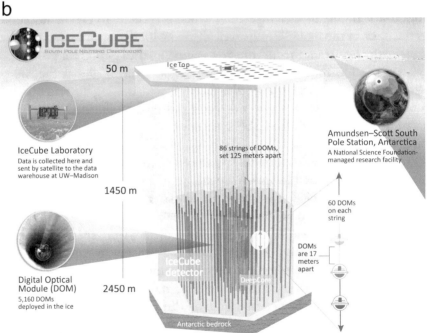

Fig. 7.11 (a) The Surface laboratories of the Ice Cube neutrino telescope in Antarctica. (b) Diagram of the arrangement of the Ice Cube detector strings and their functions. Credits: Ice Cube collaboration/NSF

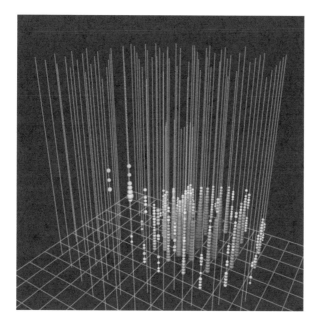

Fig. 7.12 Schematic of the detection of an ultra-high energy neutrino by the ice-cube detector. Each DOM which has detected some Cherenkov light originating in the inter-action of the high energy muon, (caused by the initial neutrino collision), with its sur-roundings is represented as a coloured spheroid. The size of the spheroid is proportional to the number of photons detected by the DOM, and the colours represent a time sequence, starting with red, through yellow and green, ending with blue. The energy and the direction of the incoming neutrino can be reconstructed from this detection pattern. Credit: Ice Cube collaboration/NSF

is a schematic showing the effects of an ultra-high energy neutrino on the matrix of DOM detectors in Ice Cube.

The most exciting observations by Ice Cube have been neutrinos of extremely high energy. The experiment was completed in 2010 and by 2013 it had detected neutrinos with energies higher than 1 PeV, (PeV stands for PetaelectronVolt i.e. 10^{15} electron volts), with asymmetries in their origins on the sky clearly pointing at direct astrophysical sources, and not sources medi-ated by high energy atmospheric cosmic ray neutrinos. But the directionalities of these sources were not adequately pinned down due to the spread of uncer-tainty in the muon track directions. But on 22nd September 2017 a high energy muon track triggered an international alert and within one minute follow-up searches were initiated with satellite and ground- based telescopes at a variety of wavelengths. On September 28th the Fermi Large Area (FermiLAT) gamma ray satellite telescope reported that the direction indi-cated was within 0.1 degrees of a catalogued gamma-ray source. This source is a "blazar" a quasar showing bursts of variable emission, catalogued as TXS 0506 + 056, and among the follow-up observations were those of the MAGIC telescopes at the Roque de los Muchachos Observatory on the Canary Island of La Palma, which detect gamma-rays from astrophysical sources from the Cherenkov light emitted by the shower of high-speed elementary particles they produce. MAGIC found gamma-ray flaring from TXS 0506 + 056 with photons of energies up to 400 GeV, and there were complementary verifications by observers at optical X-ray, and radio wavelengths. This event was notable because it was the first time since SN 1987a that an astronomical source of neutrinos had been clearly identified. The energies of the gamma rays and above all of the neutrinos show that at the centres of blazars there are mechanisms capable of accelerating particles to PeV energies, and that these sources might well be responsible for at least a fraction of the highest energy cosmic rays, which have over a million times more energy than the most energetic particles produced on Earth, in the Large Hadron Collider at the CERN laboratory in Geneva. Neutrino astronomy has aspects which make it almost paradoxical. Who, in earlier centuries, would have imagined that astronomers would be using telescopes down mines, underwater, or buried deep in the Antarctic icecap, and detecting sources not in their own hemisphere of the sky, but in the celestial antipodes!

Further Reading

Book: Neutrino. Frank Close. Oxford University Press. 2012 ISBN-10: 0199695997
A general outreach introduction by a distinguished physicist and populariser
Book: The Telescope in Ice, at the South Pole Mark Bowen. St.Martin's Press New York. 2017 (popular)
Book Solar Neutrino Physics, the interplay between Particle Physics and Astronomy Lothar Oberauer, Aldo Ianni, Aldo Serenelli, Wiley, New York. 2020 (Technical)

8

Gravitational Wave Astronomy

8.1   Gravitational Waves and Einstein

The first direct detection of gravitational waves from an astronomical source by the Laser Interferometer Gravitational Wave Observatory (LIGO) which was announced on February 11th 2016 marked a new and vital step forward in our view of the universe. Gravitational waves have been described as ripples in the fabric of space-time caused by some change in the structure of massive bodies producing a deformation of a local gravitational field, which propagate outwards from its source at the speed of light. Einstein developed his theory of gravitational interaction, General Relativity, completing it in 1915. He had, along with a number of eminent theoretical physicists, been looking at a problem first put by Newton after developing his law of gravity. Does gravity act instantly or does its effect propagate at a finite velocity? The simple question "If the Moon disappeared would the lunar tides disappear instantly, or would it take a finite time for the effect to reach the Earth" illustrates the problem. Newton had not tried to give an answer to it; it was enough to have found a law which predicted pretty well all known movements of the planets and linked them to the tendency of things to fall to the ground. But although he said to those who questioned the principles behind his law that "Hypotheses non fingo" (I do not pretend to go deeper into the underlying physics) he was never satisfied with the "action at a distance" implied. Between Newton and Einstein, a number of important theorists had looked into this and it was generally agreed that an ideal role model for theories of propagating forces was the theory of electromagnetism by Maxwell. In that theory electromagnetic waves are emitted by accelerating currents (later known to be accelerating

© Springer Nature Switzerland AG 2021
J. E. Beckman, *Multimessenger Astronomy*, Astronomers' Universe,
https://doi.org/10.1007/978-3-030-68372-6_8

electrons) and they propagate from their source at the velocity of light. A steady current produces an electromagnetic field around it, and any changes in the current produce electromagnetic waves. Maxwell produced a mathematical description of the field and the waves, which experimentalists have used to help understand, and work with, many properties of light and of other electromagnetic radiation (radio waves, X-rays, gamma rays). The question was to what extent could Maxwell's theory act as a model for gravity. This was a question for physicists, but as astronomers receive almost all of their information about the universe from electromagnetic radiation, they were amongst those most interested in the answer. Curiously Einstein himself had serious doubts about the reality of gravitational waves, even though they arise naturally within General Relativity. Henrí Poincaré had theorised about gravitational changes being propagated as waves ("ondes gravifiques") as early as 1905. However, in 1916 Einstein's theory predicted correctly the annual change in Mercury's perihelion, which had resisted Newtonian gravitation, and in a paper with approximate solutions to his basic equations introduced the concept of gravitational waves, but with an incorrect relation between the source motion and the gravitational wave strain. In a further follow-up article in 1918 he did obtain the correct formulation for gravitational waves, showing that a disturbance in the curvature of space-time produced by a change in the mass configuration of any object would propagate as a *transverse* wave distorting space-time as it travelled. This was valid but even so Einstein was not fully convinced of the reality of the phenomenon, and he even published an article in 1936, with his junior colleague Nathan Rosen, in which they thought that they had shown that the original prediction was erroneous. This was due to an error, corrected soon afterwards by Howard Robertson. Firm and general belief in the existence of gravitational waves had to await the work of Hermann Bondi, Felix Pirani, and co-theorists towards the end of the 1950s. Only since then were physicists encouraged to try to resolve the technical difficulties of detecting them.

8.2   Indirect Evidence for Gravitational Waves: Pulsars in Binary Systems

One of the problems facing those who want to test the theory of General Relativity against experiment or observation is that in most familiar circumstances its effects in "correcting" Newtonian mechanics are in general small. Relativistic effects become important where the gravitational field is strong, and this occurs particularly around compact objects, such as white dwarfs, neutron stars, and black holes. A neutron star is produced as the core of an intermediate mass star implodes after a supernova explosion, and the dramatic reduction in radius is accompanied by an equally dramatic increase in rotation velocity, as the star tries to conserve angular momentum, (the classic case of a ballet dancer drawing her arms inwards to spin faster is always used to help make this idea clearer). The result is a rapidly spinning very dense star. We meet these stars in the chapter on radioastronomy, known by their name of pulsars. They emit narrow beams of radiation at radio wavelengths, and if the beam from a pulsar crosses our line of sight, we see a radio pulse for every rotation of the star, an effect comparable to that of the rotating beam of a lighthouse. Typical spin rates for pulsars are one per second or less, going down to periods of a few milliseconds. In 1974 Russell Hulse and Joseph Taylor observed a pulsar and found its period to be 59 milliseconds. Further observations showed that the pulse interval had periodic changes, it grew and diminished again with a period of 7.75 h which was interpreted as showing it to be in a binary orbit with another star. Taylor and Hulse realised that the density of the pulsar, and of its companion (probably another neutron star but with its beam not pointing in our direction), and the orbital period showed that their movement should be subject to significant effects of general relativity. One of these is the prediction of energy loss due to gravitational waves, which should cause the stars to approach each other and their orbits shrink, until a final dramatic merger. Such is the constancy of the rotation rate of a pulsar that it can be used as a precision clock. Hulse and Taylor, (and later Taylor with other collaborators) understood quickly that it could be used for tests of general relativity, one of them being the rate of reduction of its orbital period with time.

In Fig. 8.1 you can see the results of the measurements over a period of 30 years from the discovery date. The points show the measurements, and the curve shows the prediction of General Relativity. Clearly the agreement is spectacularly good. The point here is that the loss of energy causing the orbit to decay is due to the emission of gravitational waves by the binary pair, and this is counted as the first observation showing that systems do emit gravitational waves, but of course it is an indirect inference. The rate of emission of the waves is modest, only 1.9% of the power in light emitted by the Sun. But this, and related measurements showed the validity of General Relativity, and Hulse & Taylor received a merited Nobel Physics Prize in 1993. There are now tens of known cases of pulsars in binary systems, and this is a rich field for testing details of the predictions of General Relativity and comparing them with alternative theories. So far GR has emerged as superior to its rivals. Nevertheless the validity of its prediction of gravitational waves still required direct proof.

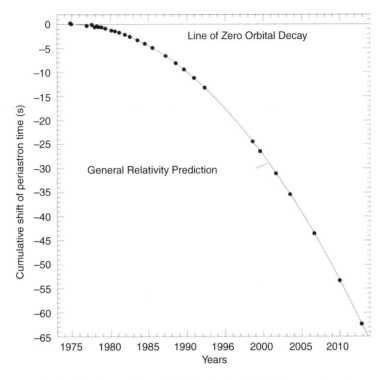

Fig. 8.1 Gradual reduction of the orbital period of the binary pulsar discovered by Hulse and Taylor (dots are observational points) superposed on the theoretically predicted graph from General Relativity, which assumes that the system emits gravitational waves. Before the first LIGO detection this was already powerful circumstantial evidence for gravitational waves. This is an extended, updated version of Fig. 2.15 Credits: J. Weisberg, Y. Huang 2016, Astrophysical Journal Vol. 289, p.55

8.3   Theorists give Reasons for the Detectability of Gravitational Waves

Physicists have been discussing gravitation in depth since the first of a series of notable conferences, at Chapel Hill, North Carolina in 1957, attended by some of the leading younger theorists of the day, Richard Feynman and Julian Schwinger, who were later to be Nobel Laureates, and John Archibald Wheeler, one of the leading gravitationalists of our time. Among the attendees was Hermann Bondi, one of the co-authors of the famous Steady State model of the Universe, which was for nearly a decade a rival of Big Bang cosmology, until shown to give predictions contrary to observation. We have mentioned him earlier as one of the theorists who nailed down the theory underpinning the prediction of gravitational waves within General Relativity. At this

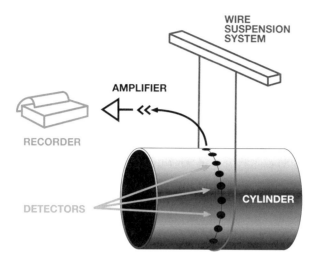

Fig. 8.2 Simplified diaram of Joseph Weber's gravitational wave antenna detector. Credits: Adapted from Crevantes-Cola et al. 2016 arXiv. 1609.09400/IAC-UC3

meeting Feynman gave a simple argument to show that a gravitational wave, though transverse, would affect a piece of apparatus by doing work on it, and convinced most of those present that gravitational waves would be, in this sense "real". One of those at the meeting was Joseph Weber, then an engineer at the University of Maryland, who decided to pit his wits against the challenge of producing an instrument which could detect the passage of gravitational waves.

8.4 Experiments to Detect Gravitational Waves

8.4.1 Joseph Weber's Experiments

Weber's basic idea was to build an "antenna" in the form of a large metal cylinder which would oscillate when a gravitational wave passed through it. He published his design in 1960, but the complexity and the difficulty of building a suspension free from other types of local oscillations entailed 5 years of construction, so the instrument was set operational in 1965. The basic element was a cylindrical aluminium block, 66 cm in diameter and 153 cm in length, weighing 3 tons. A simplified diagram of this is shown in Fig. 8.2.

The detectors convert mechanical distortions into electrical signals. The whole system protects the detector from even the smallest local vibrations, to the point that it would be able to detect reliably oscillations with the amplitude of 10^{-16} metres, smaller than the diameter of a proton, the signals expected from astronomical sources were not expected to be as big, and it was predicted that they would be lost in the noise caused simply by thermal oscillations within the system. However in 1969 Weber published an article in which he claimed to have detected gravitational waves, and in 1970 he claimed to have detected waves coming from the centre of the Milky Way. Based on the strengths of the signals he reported theoretical physicists Denis Sciama, Martin Rees, and George Field compared the amount of mass which had to be converted into energy in a given time to produce the waves with the limit that they could put on the amount of mass disappearing from the centre of the galaxy knowing that this would "unbind" it so that it would expand to a measurable degree. The limit on this expansion corresponded to 200 solar masses per year converted into energy, whereas Weber's detections would have implied over 1000 masses per year were being converted. There was considerable discussion about the results and theorist Peter Kafka calculated that a better value for the amount of mass conversion as derived from the "detected" oscillations would be 3 million solar masses per year, way over any limit set by the absence of measured expansion of the Milky Way. A good idea of Weber's was to build two detectors, separated by the 950 km between the University of Maryland, and the Argonne National Laboratory near Chicago. It was clear to him, and to his colleagues in general, that any signal detected in both detectors had to be due to gravitational waves, as no other source of mechanical vibration could act coherently over this kind of distances. During the years following these first detectors a number of experimenters constructed detectors with comparable designs, making improvements, notably cooling the cylinders to reduce thermal noise. It became clear that Weber's original results were not reproducible but although he had not detected gravitational waves he is recognised as playing a pioneering role in the field. And although none of the experiments in the 1970s was successful, the indirect detection via the measurements on the binary pulsar described above gave physicists a decisive impulse to carry on with their efforts.

8.4.2   Towards a Suitable Detector

From the early nineteen seventies a number of physicists thought that the use of an interferometer to detect gravitational waves offered the best possibilities. As early as 1962 the Russians Gershtenstein and Pusovoit had written a paper suggesting the use of interferometers, but this did not reach scientists in the West until a visit by Braginsky in the late 1970s. This chapter does not aim to give an exhaustive account of the birth and development of these techniques, and we cannot give the deserved credit to all concerned. It is reasonable, though, to pick out Rainer Weiss for his continued and effective work in seeing the virtues of the method from early on, and in envisioning the demanding experimental requirements. He encouraged Robert Forward, a former student of Joseph Weber, to build a laboratory interferometer at the Hughes Research Laboratory in California; Forward had been given the idea of using interferometry to detect gravitational waves by Joseph Weber himself. The interferometer was quite small, compared with later developments, having arms of length 8.5 metres. After 150 h of observations he reported that he had not measured any signal.

8.4.3 The Michelson Interferometer

It is clear from a detailed document by Weiss in 1972 (Weiss, R. et al., Quarterly Progress report 1972, No 105. 54–76 Research Laboratory of Electronics, MIT) that he had a conceptual design of this kind of interferometer detector worked out. His design had many of the elements which were adopted by the modern series of detectors which eventually led to the detections of gravitational waves. The basic principle is that of a Michelson interferometer, a not particularly complex optical instrument invented by Albert Michelson and Edward Morley which they used in a famous experiment in 1887 to try to measure the way our motion affects the velocity of light as we measure it.

A schematic diagram of their instrument is shown in Fig. 8.3, and consists of a monochromatic light source which illuminates a half-silvered mirror at 45° to the incoming beam. The light is split, with half reflected and half passing through this mirror. Each beam then falls onto a plane mirror which reflects the light back to the half-silvered mirror, where again half of the light is reflected and a half transmitted from each beam. The result is that, if we ignore any light lost on the way, half of the light ends up at the light source, and the other half goes to a detector. The key to the use of this device is that two monochromatic beams which combine will show interference. If one of the paths is a whole number of wavelengths (including zero) longer than the other, the two signals add together in phase, but if one of the paths is an odd number of half-wavelengths different from the other the signal at the detector is zero. The interferometer can be used, therefore, to detect small changes in the length of one or other, or both, of the distances of the plane mirrors from the half-silvered mirror. The mirrors are spaced to produce the zero signal, and a

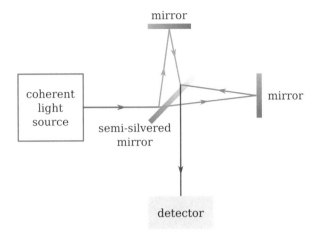

Fig. 8.3 A schematic diagram of a Michelson Interferometer, used in the historical Michelson-Morley experiment to measure whether the velocity of light varies with direction in space. The negative result was a basis for Einstein's Special Theory of Relativity. Credit: Benjamin D. Esham, Public Domain via Wikipedia Commons

small change will yield a finite light signal at the detector, which can be used to measure the size of the change.

In the Michelson–Morley experiment it was not the distances of the mirrors which were being tested, but the speed of light. If there is a difference in the speed of light between the two paths, the result on the interference pattern is the same as changing the arm lengths. The idea of the experiment was to see if the Earth's velocity through a supposed stationary universal "ether" would cause a change in the interference pattern. We know that the Earth moves in its orbit round the Sun at a speed of some 30 km/sec, so if this speed affects our measurement of the speed of light, this should show up by making measurements as we move around the orbit. Michelson and Morley found no measurable effect of this kind, and their experiment has been repeated with increasing precision and a variety of methods since then, with the same result, there is no measured change in the speed of light when measured at any time or place and with the apparatus moving at any speed. This result was the basis on which Einstein produced his Special Theory of Relativity in 1905, with all of its consequences for the development of modern physics. In these experiments the two perpendicular arms of the interferometer had lengths of a few metres, ranging from 2 to 32. We will see that for effective detection of gravitational waves the interferometer arms have to be orders of magnitude longer.

8.4.4 The LIGO Concept

The basic plan whose development led to LIGO is based on four principal components: a Michelson interferometer using the interference principle to detect minute changes in the length of its perpendicular arms caused by the passage of a gravitational wave, a laser light source which allows the interference to be measured with the highest precision, an optical device which boosts the effective power of the laser, and special optical elements in each of the arms to give them a much longer effective optical length. These optical elements are modifications of pairs of parallel closely spaced partially reflecting transparent plates, originally called Fabry-Perot (FP) etalons, after the French optical physicists Charles Fabry and Alfred Perot who devised them in 1859. They are now widely used in telecommunications, often coupled to fibre optical systems. In LIGO the role of these modified FP systems is to produce multiple reflections along the arms, thereby lengthening their optical length. They work on an interference principle and their use allows the light to be shuttled back and forth along the arms with extremely low light loss at each reflection. It was recognised by those who devised LIGO that a gravitational wave will cause a change in the length of a test rod which is proportional to its own length. So in order to have measurable changes in length you need a very long rod, much longer than the lengths of the aluminium blocks in Weber's experiments. Although it is not practical to make a rigid kilometre long rod, this can be overcome using the optics of the Michelson interferometer, where instead of a rod you have the distance between the central optics and the reflecting mirrors at the ends of the arms. In LIGO these arms are 4 km long. But the use of the sets of FP double plates increases the effective optical path by the number of times the light is reflected, which for LIGO is 280 times, thus giving an effective path length of 1120 km! The laser boosting optics is called a "recycling mirror", which is situated between the laser and the 45° beamsplitter. This is also partially reflecting, so that the light which would, after traversing the interferometer arms, return to the laser is intercepted and reflected back into the interferometer. For LIGO the highest input laser power used so far is 50 watts, and the typical gain produced by recycling is around 60 so the effective power In the Michelson is some 3 kilowatts. The power gain produced by the use of the FP's is about 80, so the effective power in the FP cavities is 240 kilowatts. This whole arrangement is schematised in Fig. 8.4, where the positional distribution of the optical elements is shown. Using this figure as a reference the passage of a gravitational wave causes a strain in space, measured by the ratio of the change in length ΔL in a path of

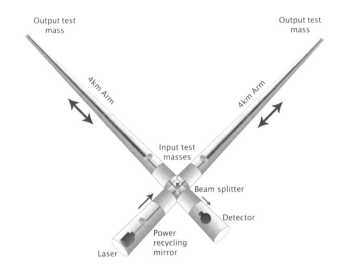

Fig. 8.4 Diagram of the basic layout of the optics used in LIGO. Light from the input laser is split at the beam splitter, placed at 45° to the input beam. The transmitted and the reflected beams go respectively to the two perpendicular arms from where, with multiple reflections and boosting (not shown) the beams finally recombine at the photodetector. A tiny change in the relative arm lengths causes a change in the strength of the light signal reaching the photodetector. Credit: Chris North, Cardiff University

length L which is positive in one direction but negative in the perpendicular direction; both of these directions are themselves perpendicular to the propagation of the beam travelling along an interferometer arm.

8.4.5   Projects Which Tested the Use of Laser Interferometers

The concepts which went into LIGO were developed in a collaborative way over several decades, and I will summarise only very briefly some of the more important steps, to give an idea of how complex a project it was, but without aiming for a full history, for which the reader should consult work devoted entirely to LIGO. An original concept at MIT by Weiss' group was embodied in a report, the RLE QPR report, in which many of the basic elements of the system were outlined: multi-pass optical delay lines, pendulum suspensions, servo systems to hold the interferometers steady on a single light fringe with radio frequency modulation etc. This was first carried into practice by a group under Heinz Billings, at the Max Planck Institute for Quantum Optics (MPQ) at Garching, Germany, who after seeing the report asked Weiss if they could build a trial system based on its principles. He agreed, and they built a

1.5 m prototype, which served to identify a problem requiring laser stabilisation which he had not foreseen. Solving this, and building a successful 3 m system in which all the optics in the phase sensitive parts were on suspensions, (see Figs. 8.6 and 8.7) they upscaled to a 30 m instrument, which had better strain sensitivity than any previous bar detectors. The success of the 3 m instrument encouraged Weiss to produce the "Blue Book" ("A study of a long baseline Gravitational Wave Antenna System") produced for the National Science Foundation, coauthored by Peter Saulson and Paul Linsay, as well as by Stan Whitcomb at Caltech, and submitted to the NSF in 1983. In the meantime Ron Drever at the University of Glasgow had introduced optical storage elements using Fabry-Perot cavities in his 10 m gravitational wave detector. He moved to Caltech and built a 40 metre prototype there, in which several new ideas were developed. He invented an elegant technique to stabilise the laser frequency, as well as the idea of power recycling, which was developed independently by Schilling in the Max Planck group. The idea of signal recycling was invented by Brian Meers, at Glasgow.

In the 1980s both the Glasgow group and the Garching group made proposals for the ambitious long baseline interferometers which they knew were needed to finally observe the long sought gravitational waves, but in both cases their funding authorities did not give the projects priority, and they were not funded. But a joint British-German proposal for a reduced scale, 600 m instrument was funded from 1995, it was built 20 km south of Hanover, and named GEO. It started operation in 2002, and served as a test bed for the techniques described above, in particular the power recycling technique, originally suggested by Ron Drever, was tested at GEO.

In parallel with these developments French and Italian groups led by Allain Brillet and Adalberto Giazotto, respectively, were working to promote an instrument for gravitational wave detection. Giazotto, in particular, was developing attenuators for isolating the suspended mirrors in an interferometer from seismic vibrations. The groups, originally from Orsay and Pisa, and later joined by others at Frascati and Naples, applied for funding from the French CNRS, which was approved in 1993 and the Italian INFN in 1994. A site near PISA was chosen for the project, which was named Virgo. Organisation and construction began in the late 1990s and the initial VIRGO detector was completed in 2003. From 2007 Virgo and LIGO collaborated formally in the search for gravitational waves, but fluid informal collaboration between them, GEO, and the MIT gravitational wave laboratory had begun in the 1980s and interchange of staff and graduate students were continuous throughout the development of the different detectors.

8.5   The LIGO Project: Conception and Implementation

LIGO, the most successful project, the one which made the first detections of gravitational waves, was a joint enterprise between MIT and the California Institute of Technology (Caltech). The leading scientific roles were played by Rainer Weiss at MIT and the theoretical astrophysicist Kip Thorne at Caltech, where Ron Drever was invited to work on gravitation wave techniques. After the presentation of the Blue Book mentioned above the joint Caltech-MIT project was funded by the NSF, but progress was quite slow, until Barry Barish, who had considerable experience in managing projects in high energy physics, took over as project manager in 1994. He renewed all planning aspects, and set as a first goal the development of "initial LIGO" (iLIGO) in which the concept would be rigorously tested, and which could possibly detect gravitational waves, followed by a second stage of "advanced LIGO" (aLIGO) in which detection possibilities would be enhanced. The project entailed two widely separated observatories, one of which was built at Hanford, in Washington state, and the other at Livingston, Louisiana, and these were constructed between 1994 and 1997. Barish then separated the management of the laboratories and a science team (LIGO Scientific Collaboration, LSC,) headed by Weiss, which built up collaborations with scientists in general, (including engineers and graduate students) and with Virgo and GEO in particular. Jts first spokesperson was Weiss. After that the position was based on elections, and has subsequently been taken successively by Peter Saulson, David Reitze, Gabriela Gonzalez, David Shoemaker, and Patrick Brady. The LSC came to involve over 1000 scientists and engineers, so LIGO has taken its place among the "big science" projects typified by major particle accelerators, huge modern telescopes, and satellite observatories. Its primary function has been to perform the data analysis and write up the scientific results. It has committees on future directions for the instruments, and on the astrophysical interpretation of the results.

iLIGO operated between 2002 and 2010 without detecting gravitational waves. It had reached its design sensitivity apart from a frequency range below 30 Hz. The upgrade to aLIGO took place between 2010 and 2015, and included technical contributions developed from 2005 onwards from a wide variety of projects including VIRGO and GEO. The LIGO laboratory at Stanford University contributed to active vibration oscillation for the optical components GEO and Glasgow University made advances in fused silica suspensions and mult-element suspensions, deterministic locking techniques were developed both within LIGO itself and in Australia. The University of Florida at Gainesville worked on phase modulators and Faraday isolators, and later both GEO and Australia supplied "squeezed" light sources. If you are interested in the technical details of all of these you will be able to read about them in a unique technical history of the project in preparation by some of its leading participants, which should be published some time during 2021, but here I want to give just a flavour of the extent and depth of the work involved.

8.5.1 Some Details of LIGO

8.5.1.1 The Optical Beam Paths

We can see the two sites in the aerial views in Fig. 8.5a, b.

The layout is straightforward and the same for both sites, a central laboratory, and from it extending the two arms of the interferometer, each of length 4 km. The tubes must be evacuated fully along their lengths, to one million millionth of an atmosphere. This was established at the start by baking the tubes at 150 celsius for a month, and is maintained by ion pumps at the ends and the middle of each tube. The tubes are closed off by gate valces when work is done on the detector components. This practice avoids the need for repeating the baking process, and the tubes have not been exposed to atmosphere since the bake-out. The vacuum is needed to reduce the noise from fluctuations in the forward scattering of the laser beam by residual gas molecules, and also to reduce the chance of hydrocarbons sticking to the mirrors reducing their reflectivity. To ensure a perfectly straight path, GPS-assisted earth moving and concrete support installations were necessary to counteract the curvature of the Earth, which gives a deviation of some 30 cm over a 4 km arm. The beam tubes are of highly stable rust-free steel; their separate sections were welded together with a continuous spiral weld, and they have reinforcing rings at short intervals to prevent collapse.

8.5.1.2 Stability of the Optics

The full optical cavity comprises an end test mass, ETM, (mirror) with a high reflectivity coating facing the input test mass, ITM, (mirror) which has a small transmission coating facing the ETM. Both ETM and ITM have reaction masses in close proximity. The reaction masses have electrostatic electrodes to control the positions of the ITM and ETM. A key element in the

a

b

Fig. 8.5 (a) The Hanford site of the LIGO experiment. The two perpendicular arms for the light paths are seen emerging from the central laboratory buildings. (b) The central laboratory complex of the LIGO site at Livingston; the interferometer arms can be seen emerging at right angles to one another. Credits: Caltech/MIT/LIGO Laboratory

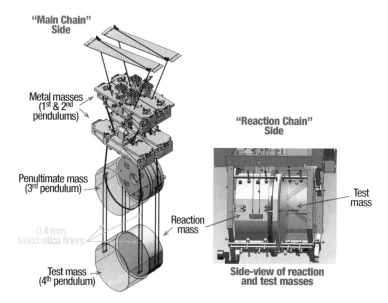

Fig. 8.6 Diagram of one of the suspension systems for the test masses which carry the reflecting mirrors at the ends of the 4 km evacuated tubes. The complexity is due to the essential need to damp out all local sources of vibration. Credits: Caltech/MIT/LIGO Laboratory

suspensions are fused silica fibres that support the test masses from a penultimate mass. The silica fibres have very low mechanical loss, and thereby introduce very little thermal (Brownian) noise to the test masses. The rest of the structure of the four masses suspended in series is designed to reduce the vibrations of the ground on the test masses. The reaction masses are made of fused silica as they need to be transparent to the laser beams. They are suspended from independent chains of masses. Elements of these structures are shown in Fig. 8.6.

The optical test masses are held in place against metal reaction masses by an exquisitely balanced system of electromagnetic and electrostatic forces. It is necessary to maintain the optical distances between the test masses constant to within less than a thousand millionth of a meter. Figure 8.7a shows one of the Hanford test mass suspensions, during a laboratory test, while in Fig. 8.7b we see two of the fused silica end mirrors, each weighing 40 kg.

The suspended optical systems are isolated by an active isolation protection system, employing position and vibration sensors tuned to various frequencies of environmental vibrations, together with permanent magnet actuators

a

b

Fig. 8.7 (a) One of the Hanford test masses under laboratory test. (b) The fused silica end mirrors at Hanford. Credits: Caltech/MIT/LIGO Laboratory

working together to counteract ground movements. Seismometers and accel-
eromators are mounted in the isolation system as reference elements to measure the motion of a platform. The platform itself has actuators mounted on
it to counteract the motion of the platform. The actuators reduce the motions
as measured by the reference elements – making a feedback system that
reduces the vibrations of the platform induced by external vibrations. The net
result of the multiple suspension system supported by the active seismic isolation systemis to reduce the external vibrational noise at 100 Hz from 10^{-10} to
10^{-19} metres (a ten thousandth of the effective radius of a proton) at the
test mass.

8.5.1.3   The Laser and the Optical Path

The laser which produces the light whose interference allows the detection
of gravitational waves operates at the near infrared wavelength of 1.06
microns. A "seed" beam of only 2 watts of power is progressively amplified
in stages passing it through optically active material until it reaches 50
watts. It is passed through a feedback system which reduces its instabilities
in frequency and power by a factor of 10^8 before passing into the input
optical system. Passing through a system of etalons it reaches the 45° beamsplitter of the interferometer boosted to an effective input power of 3 kW,
from there the reflected and transmitted beams go through their respective
intermediate test masses (ITM's) which contain Fabry-Perot cavities achieving an effective power of 250 kW in each beam. After reflections at the end
test masses (ETM's) the two beams pass back through the ITM's and converge again on the beamsplitter. There the carrier light from the two arms
interferes, so that it all goes back to the recycling mirror, while the gravitational wave signal (the sidebands) carried by light from both mirrors is
added and sent to the detector. A tiny fraction of carrier is deliberately
leaked to the detector as well, to make the signal detectable. This is the way
to describe the interferometer, when a servosystem has locked the interferometer to a dark a fringe at the photodetector. An additional datum, also
for those used to dealing with signals, the frequency noise at 100 Herz
measured at the detector is 5×10^{-7} Hz/(Hz$^{1/2}$) and the relative amplitude
noise is 10^{-8} Hz$^{-1/2}$. A scheme of the optical paths described in this para-
graph is given in Fig. 8.8. For those who want to pursue details of this sort
for a better understanding, you should follow up on the reading list for this
chapter.

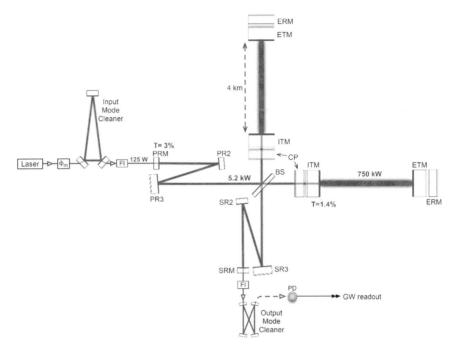

Fig. 8.8 Diagramatic representation of the Advanced Ligo Optical system Credit: LIGO Scientific Collaboration

8.6   The First Detection of Gravitational Waves

8.6.1   The Observation

Gravitational waves were first detected by LIGO on September 14th 2015. The signal passed through the two detectors with a separation in time of 7 milliseconds. This occurred during final commissioning tests just 4 days before aLIGO was due to make its first official run. The now famous signals, detected at Hanford and Livingston, are shown in Fig. 8.9.

What we can see in the top part of the figure are the two signals just as they were received. On the left is the Hanford signal, in red, and on the right the Livingston signal, in blue, shifted by the time interval, and superposed on the Hanford signal, showing their strong similarity. The only known physical effect which can produce this simultaneous oscillatory change in dimensions in two pieces of apparatus separated by 3002 kilometres is the passage of gravitational waves. In the panels below are comparisons with the theoretical predictions, which we will mention shortly. We can see that, apart from minor

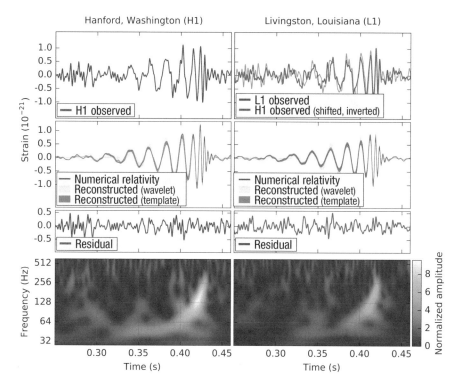

Fig. 8.9 The first gravitational wave, detected by aLIGO on September 14th 2015. Upper panel left, the Hanford signal, upper panel right, the Livinston signal shifted by 7 milliseconds, the time difference between the detections at the two sites. Upper middle panel, comparisons of the two signals with a theoretical model , lower middle panel residual differences between the signals and the model. Lower panel, the frequency spectra of the two signals. Credit: LIGO Scientific collaboration and Virgo collaboration

local noise effects, the two wave trains are identical, and taken together the observers talk about the detection as a 5σ detection, which is highly statistically significant; normally detections with uncertainties of 3σ or even 2σ are considered reasonable evidence, but in a case as important as this, it was as well that the detection left virtually no room for doubt. In the lowest panels are plotted, for each of the detectors, the frequency spectrum as a function of time. We can see that the whole detectable event lasted barely one tenth of a second, and that the frequencies of the waves started low, at some 35 Hz and rapidly shot up to over 150 Hz. We can also see this in the upper curves as the waves compress with time. The form of this signal is nicknamed a "chirp", and indeed its frequencies lie within the audible range, although the power detected was minute. The source, once confirmed, was labelled GW150914.

8.6.2 The Interpretation

While the experimentalists had been working over the years on the extreme demands of the experiment, the theorists had been building up their predictive powers for sources which could emit gravitational waves with amplitudes sufficient to be detected. Already from the 1980s, with the discovery of pulsars in binaries, and the observation of the decay showing that gravitational waves must be emitted, a series of articles had given predictions of the characteristics of detectable events. In a comprehensive review article in 2014 Luc Blanchet set out the theoretical framework for analysing the results of gravitational wave observations specifically aimed at handling the phenomenon of the inspiralling and merger of two highly compact objects, which is predicted to be one of the most common types of generic ways in which recognisable signatures of gravitational waves are produced. These compact bodies include principally black holes and neutron stars, because although white dwarfs are compact, they are less compact, and less massive than the other two types of stars. All three of these types of objects are expected to be relatively common because they are the end products of normal stars. There are much more massive black holes in the centres of galaxies too, and these should also be capable, under circumstances, of producing gravitational waves, but for statistical reasons compact binary stars were considered much more likely to be detected, as indeed has proved to be the case. The article related the parameters of the emitted wave to the properties of the inspiralling binary pair, notably their masses, their angular momenta, the progressive reduction in their orbits as they spiral in towards a merger, and the energies emitted per unit time. This and a considerable range of referenced studies, enabled the observers to infer an immediate estimate of the total mass of the pair of objects from the lower frequency range of the observed chirp, with a value of 70 solar masses. They could also immediately determine the sum of the "outer" radii of the two components if they were black holes, (the Schwarzschild radius), with a minimum1 value of 210 km. To attain an orbital frequency of 75 Hz (i.e. 75 orbits/second) which is half the upper value of the gravitational wave frequency, the objects had to be very close together. Two equal mass "point sized" objects would have to be some 350 km apart to be rotating around each other so quickly. So the objects had to be very compact to still be orbiting without contact. Only black holes or neutron stars are compact enough to satisfy this condition. But a neutron star binary would have emitted far less energy, and a black hole with a neutron star companion would have implied a much more massive black hole, giving a much lower merging frequency. So the deduction was that the gravitational waves emitted by GW150914 came from the merger of two black holes of similar masses, both close to 30 times the mass of the Sun. After making this basic inference, the theorists could then compute the theoretical waveform for the emission, and also for the final oscillations of the single black hole resulting from the merger, which also match the shape of the waveforms after the maximum, which agreed very well with the forms observed, as displayed in some detail in Fig. 8.10. Computing the predicted power output in gravitational waves enabled the LIGO team to calculate the distance of the event at 410 Megaparsecs, some 1300 million light years.

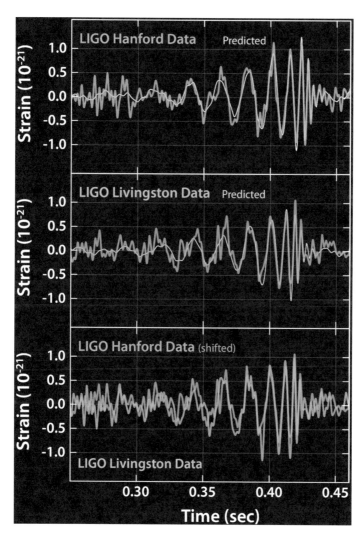

Fig. 8.10 Detailed comparisons between observation and prediction for the gravitational waves detected from source GW150914. Credit: LIGO Scientific Collaboration

This was a dramatic first detection. Not only was it indeed the first time gravitational waves had been detected, but it also showed the reality of black holes in a high stellar mass range, significantly higher than had been detected so far for stellar mass black holes in the Milky Way. We have met these latter in the chapter on X-ray astronomy, as they are found by detecting X-ray emission from binary stars, where one of the two objects is a black hole, and the other is a "normal" star feeding it. The Milky Way stellar black holes have masses around 10 solar masses, in a range from some 4 to a maximum of 14, but the first black holes detected via the gravitational waves just before their merger had masses more than twice this upper value. There was a missing element in the detection. As the wave train was seen by only two detectors, it was not possible to locate the direction of the source in the sky with a reasonable degree of accuracy. Another problem was that the two LIGO detectors are oriented to detect the same polarisation of gravitational waves, so information about the polarisation of GW150914 was lacking. This type of information had to wait until Virgo became operational.

8.7   The Stream of GW Detections Begins

The importance of the first detection, the long and complex development process, and the extreme sensitivity implied, made the LIGO team naturally cautious about reporting their work. The detection was made in September 2015, but by the time the work was published there were two further detections, one tentative, which was finally declared verified in 2019, and another at a satisfactory signal to noise level. These detections were sufficient for the physics community to agree that the results were in no way spurious and not a "flash in the pan" but marked the opening of a new and fundamental line of research in physics and astronomy. The importance of this was fully recognised by the award of the Nobel Prize in Physics 2017 to Rainer Weiss, Kip Thorne, who as a theorist had been working hand in hand with Weiss to pilot and sustain the LIGO project, and Barry Barish, the physicist who took over the management of the project in 1994 and provided the administrative leadership needed for the long route to success. We see them in Fig. 8.11.

As I start to write this chapter of the book LIGO, having had 2 years of down time for refurbishment, and having operated again for a further 2 years, is now being upgraded again. During its active phases it obtained a dozen detections or potential detections of gravitational waves, and these are listed in Table 8.1. The third column in the Table is interesting. It shows the area on

Fig. 8.11 From left to right, Rainer Weiss, Kip Thorne, and Barry Barish, who were awarded the Nobel Prize for Physics in 2017 for the detection of gravitational waves by LIGO. Weiss was the driver of the experiment, Thorne was the leading theorist involved, and Barish showed the administrative guidance essential in bringing the complex system to fruition. Credits: LÍGO Scientific Collaboration

the sky within which the source of the gravitational waves could be placed. The distance between the Livingston and the Hanford detectors, which causes a time difference of a few milliseconds in the arrival of a gravitational wave event, allows only a very crude placement of the source in the sky, as shown by these areas in square degrees. A thousand square degrees, which is the equivalent of an area of just over 30 × 30 degrees does give a rough fix, but conventional astronomers, using optical and radio techniques, are accustomed to fixing their new sources to within a few arcseconds, and this usually gives the astronomer only one, or at the most, two or three likely objects within the circle of uncertainty. It is at this point that the existence of the Virgo detector near Pisa was shown to have its importance. VIRGO was being upgraded when LIGO detected GW150914, but came back into service in 2017. The first event which Virgo detected was GW170814. The detection signals from the two LIGO detectors and the VIRGO detector are shown in Fig. 8.12. In Fig. 8.13 you can see the signals from all the detections until the end of 2017.

8.7.1   GW170817 A Source with Over 4000 Messengers

But the most significant intervention so far by VIRGO took place only 3 days later. A signal was detected by VIRGO and by the two LIGO detectors, which had a different scale of frequencies from those previously observed, and lasted much longer. The signals as detected can be seen in Fig. 8.14.

Table 8.1 Basic details of the detection of gravitations wave events detected before the end of 2018

GW event[a]	Date published[b]	Limiting area[c]	Distance[d]	Energy[e]	Primary[f]	Secondary[g]	Remnant[h]
GW150914	11/02/16	179	430	3.1	BH 35.6	BH 30.6	BH 63.1
GW151012	15/06/16	1555	1060	1.5	BH 23.3	BH 13.6	BH 35.7
GW151226	15/06/16	1033	440	1.0	BH 13.7	BH 7.7	BH 20.5
GW170104	01/06/17	924	960	2.2	BH 31.0	BH 20.1	BH 49.1
GW170608	16/11/17	396	320	0.9	BH 10.9	BH 7.6	BH 17.8
GW170729	30/11/18	1033	2750	4.8	BH 50.6	BH 34.3	BH 80.3
GW170809	30/11/18	340	990	2.7	BH 35.2	BH 23.8	BH 56.4
GW170814	27/09/17	87	580	2.7	BH 30.7	BH 25.3	BH 53.4
GW170817	16/10/17	16	40	>0.04	NS 1.46	NS 1.27	NS < 2.8
GW170818	30/11/18	39	1020	2.7	BH 35.5	BH 26.8	BH 59.8
GW170823	30/11/18	1651	1850	3.3	BH 39.6	BH 29.4	BH 65.6

[a]Standard reference index to the event observed; GW for gravitational wave source, followed by the detection date (year, month, day)
[b]Date of publication of the detection
[c]Area on the sky which limited the possible position of the source in square degrees)
[d]Distance in Megaparsec (Mpc) 1 parsec = 3.26 light years
[e]Energy emitted by gravitational waves during the merger (in units of solar masses divided by c2)
[f]Object type and mass of the primary (BH means black hole, NS means neutron star, mass in solar masses)
[g]Object type and mass of the secondary
[h]Object type and mass of the merged remnant in solar masses

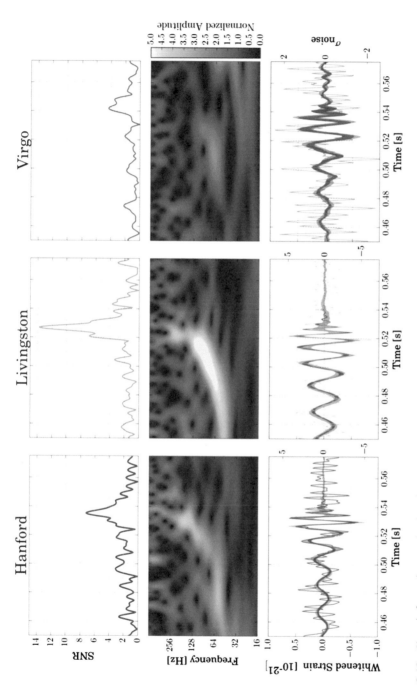

Fig. 8.12 Signals from the gravitational wave source GW170814 detected by the two LIGO detectors and also by the Virgo detector. The availability of the time differences between the three detections enabled the position of the source on the sky to be fixed within a few degrees. Credit: LIGO Scientific collaboration and Virgo collaboration

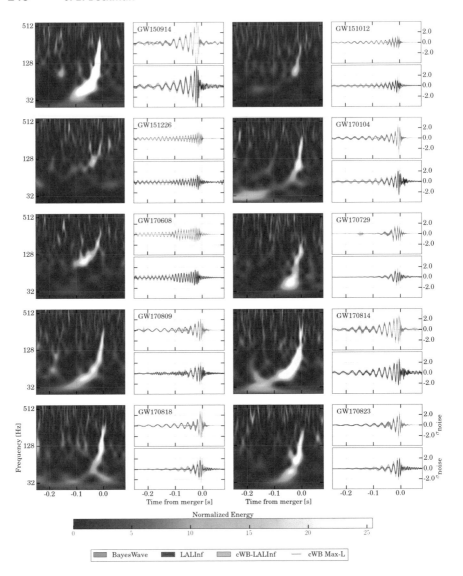

Fig. 8.13 The "chirp" signals: detections of gravitational waves by LIGO prior to its going into a recess in 2017 for upgrading, as presented in the GWTC-1 catalog. Credit: LIGO Scientific Collaboration and Virgo collaboration

Instead of a duration of order one tenth of a second this signal was detectable for more than 100 seconds, and the relatively slow process showed that the event was a merger of two neutron stars, as opposed to all the other detections of pairs of black holes. The GW signal started at a frequency of 24 Hz and increased in both frequency, to several 100 Hz, and amplitude, with a

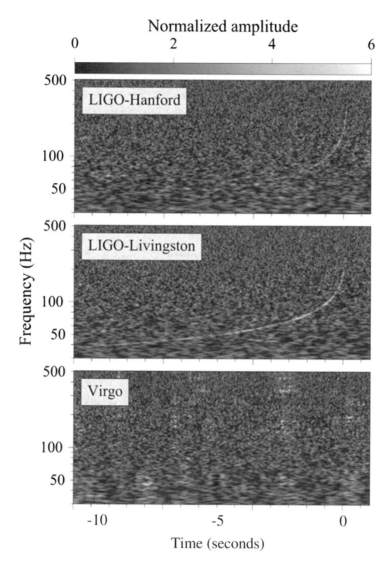

Fig. 8.14 Gravitational wave signals at LIGO from two inspiralling and merging neutron stars. This was the first signal received from an event other than the merging of two stellar mass black holes. You can see that Virgo did not receive a signal. But this non-detection was in fact very valueable, and combined with the signals from LIGO enabled the observers to place it within a 40 square degree box on the sky (see the text for the explanation) thus allowing the identification of the site of the merger, the galaxy NGC 4993. Credit: LIGO Scientific Collaboration and Virgo collaboration

wavetrain of over 3000 cycles ending in a typical chirp. The non-detection with Virgo was in fact a very significant result. Just 3 days previously VIRGO had shown that it could detect a gravitational wave source. The fact that it did not detect GW170817 implied that this source had to be in a zone on the sky within the 45 degree cone of silence perpendicular to the interferometer plane. Combining this information with the area on the sky within which the LIGO signals placed the source confined the site of the event to a box of size only 40 square degrees, 3 times smaller than would have been possible from the LIGO data alone. This was a significantly smaller error box than those from any of the gamma-ray satellite measurements which took place at the same time, and was crucial for the first optical measurements, made some 10 h after the GW and gamma-ray measurements, which showed that the signal had come from the galaxy NGC4993, relatively local at only 130 million light years from us. The detection of this GW source turned out to be the most exciting of all for the astronomers, because unlike black hole mergers neutron star mergers emit not only gravitational waves, but also a significant pulse of electromagnetic waves. Independently of the LIGO-VIRGO observation, the Fermi and the INTEGRAL gamma ray satellites detected a short, 2 second gamma-ray burst starting 1.7 seconds after the GW merger signal. Although the directionality of these satellites is not very fine, the direction on the sky indicated for the burst did overlap the GW position. It had long been predicted by theorists that short gamma-ray bursts could be caused by neutron star mergers, and this event was a direct detection and proof. The Fermi satellite detection triggered urgent observations with telescopes at all wavelengths around the world from which the source position, in the southern sky, was observable. Robotic telescopes which are very agile and designed for follow-ups of this kind were among the first to take images of the source and six teams had images including the source area within 90 minutes of the trigger. The first to detect optical light from the source was the Swope Supernova survey, with a near infrared image using the 1 m Swope telescope at Las Campanas Observatory in Chile, 10 h and 52 min after the GW detections. By a nice coincidence it was the telescope operator at the Swope, Oscar Duhalde, who with Ian Shelton first spotted supernova SN 1987a in the Large Magellanic Cloud, from which the first supernova neutrino detections were made. By identifying the source galaxy, the optical astronomers could also know the distance to the source, finding good agreement with the distance found directly from the GW signal. Once a precise position had been found optically, large optical telescopes as well as radiotelescopes, and satellites at other wavelengths could be focused on the object. 15.3 h after the trigger it was detected in the ultraviolet by the Swift gamma-ray burst satellite, 9 days later it was detected in X-rays by the Chandra X-ray observatory, and 16 days later it was detected at radio wavelengths by the Karl Jansky Very Large Array. Follow up spectroscopy from large ground-based telescopes found neutron-rich elements in a gas cloud expanding at one tenth of the speed of light. These elements, such as gold and platinum, would be expected from neutron star mergers. GW170817 was one of the most observed astronomical events in history. The leading journals, The Astrophysical Journal Letters, Science, and Nature, had articles by over 4000 authors in over 900 institutions dealing with all aspects of the merger.

Among the most interesting results of the detection of GW170817 was the possibility to show whether gravitational waves propagate at the speed of light, as predicted within General Relativity. The assumption has to be made that the first photons were emitted between the time of the peak of the gravitational wave emission and 10 seconds later. With this assumption the difference in the velocity between the gravitational and electromagnetic waves, expressed as a fraction of the speed of light, has to be between -3×10^{-15} and $+7 \times 10^{16}$. In other words, gravitational waves travel at the speed of light with a fractional uncertainty which is very small indeed. The signal from GW170817 also enabled the physicists to exclude some of the postulated alternatives to General Relativity, which for the time being remains the best description of how gravity acts.

A further deduction of interest was the pioneering use of sources of gravitational waves to measure large astronomical distances. Traditional methods using electromagnetic radiation rely on "standard candles" which are objects whose absolute luminosities, (in watts, for example) have been derived on the basis of long-term observation and theory. Starting from purely trigonometric measurements of neighbouring stars to calibrate the nearest of these standard candles, and stepping outwards with a series of increasingly luminous objects, astronomers have quantified the expansion of the universe by plotting the distances thus measured against the velocities of the objects, ascertained using the Doppler effect, the redshift of spectral lines which gives a direct velocity reading. Local standard candles include, notably, the Cepheid variable stars, whose periods are proportional to their luminosities, and the most powerful generally used standard candles are type Ia supernovae, which led to the inference that the universe is accelerating. One of the key parameters in these measurements is the Hubble-Lemaître parameter, which is just the velocity of recession of a galaxy, or a source within a galaxy, divided by the distance to the source. One of the key projects for which the Hubble Space Telescope was designed was to measure the Hubble-Lemaître parameter using Cepheid variable stars as standard candles. The values, based on galaxies nearby and in the middle distance, centred around 72 km/sec/Megaparsec. But other methods based on the properties of the cosmic microwave background give values centred on 67 km/sec/Mpc. The difference seems quite small, but it is bigger than a value which can be accounted for by observational error, and at the time of writing this amounted to a mini-crisis in observational cosmology. The inference of the distance to a source of gravitational waves does not depend on models of the universe. It is found by comparing the properties of the observed wave train with those predicted by models, to give the absolute luminosity in gravitational waves. Comparing this with the received power gives a direct measure of the distance. In the case of the neutron star merger GW170817 this yielded a distance to the galaxy NGC 4993 and a measure of the Hubble-Lemaître parameter of 70 km/sec/Mpc. As more sources are measured it will become possible to measure this parameter to accuracies of order 1%, and this will be valuable in distinguishing between cosmological models, above all to see if the standard model with its accelerating universe has any basic flaws. Gravitational wave sources, in this context, have been nicknamed "standard sirens".

A final bonus from GW 170817 has been a contribution to our knowledge of the origin of some of the chemical elements. In Chap. 1 we saw that hydrogen and most of the helium in the universe were formed in the first minutes after the Big Bang, that lithium, beryllium and boron have been produced in interstellar space from the splitting of carbon, oxygen, and nitrogen by cosmic rays, and the remaining elements, from carbon upwards in mass, have been formed in stellar interiors. Many of them are known to have been produced in supernova explosions, but the theory of production of certain heavy elements rich in neutrons does not work under the conditions prevalent in the stars which explode as supernovae. Half of the elements heavier than iron need a process called rapid neutron capture to form, and theorists had suggested merging neutron stars as possible sites for this. Computer modelling of the GW 170817 neutron star merger predicted that strontium, one of the neutron rich elements in question, should have been produced, and have left its imprint on the optical spectrum in a pair of absorption lines at 350 and 850 nanometres wavelength. Some days after the original event, the optical phenomenon of the "kilonova" was detected at the site, and observers took spectra, looking, *inter alia*, for the strontium lines. These were indeed detected, and in an article by Darach Watson and coworkers in 2019 the estimate of the amount of strontium produced was shown to be some five times the mass of the Earth. This was strong confirmation of the hypothesis that neutron stars are sites of neutron rich heavy element formation, and in a way also confirmed the nature of the merging objects in GW 170817 as neutron stars.

8.8   The Future of Gravitational Wave Research

The use of gravitational waves to explore the universe is clearly still in its infancy. It was essential to have two detectors at widely spaced sites to be able to make any valid claim for a detection. We have seen how the addition of a third site enables the location on the sky to be much better determined using the delay times between the arrival of the wave front at the three detectors. There is no doubt that increasing the number and the spacing of the detectors will enhance the capability of localising the sources of the waves. But the increase also means that different teams can try different approaches to improving the detectivity. The intrinsic sensitivity needed has been expressed by comparing the tiny fractional movement in the detected waves to the measurement of a change in the distance of the nearest star by the width of a human hair, truly a technical achievement matching the intellectual achievement of General Relativity itself. A further benefit which will arise from the use of several detectors is the possibility to measure the polarisation of the waves. Unlike electromagnetic waves which polarise in planes separated by 90 degrees, gravitational waves polarise in planes separated by 45 degrees, and measurements of the degree of polarisation can be obtained only with three or more separate and reliable detectors.

As well as future plans to keep upgrading LIGO and Virgo there are projects under way to build detectors at sites in different continents. One of these is LIGIO-INDIA ("INDIGO"). With collaborative agreement between the US National Science Foundation and the Indian Government a new detector originally designed to be placed at Hanford will be sited close to Aundha Nagnath in India. The project has been funded and a team of scientists is in training, combining local personnel and high level physicists from India to be trained by the experience staff of LIGO. INDIGO will form a part of the LIGO-Virgo international group.

An additional project designed to meet the challenge of gravitational wave detection is KAGRA. This is to be sited at the Kamioka mine in Japan, already distinguished for the Kamiokande series of neutrino detectors, discussed in another chapter of this book. The benefit of building KAGRA in a mine is the prospect that Newtonian gravitational gradients caused by the fluctuations in density of the ground due to seismic waves, (also in the atmosphere) will be much smaller than on the surface of the Earth. The noise due to these effects becomes dominant in the interferometers on the surface at frequencies lower than about 5 Hz. KAGRA should bridge the gap between the very low frequency sources expected to be detected by LISA (see just below) and the current GW detectors. The KAGRA range should detect the type of sources detected by LIGO and VIRGO but at a prior phase when their signals are weaker and at lower frequency. It should also be good for detecting higher mass black hole mergers.

The Japanese physics community has been working on gravitational wave detection since the 1990s with prototype detectors TAMA 300 and CLIO operating respectively from 1998 to 2000 and since 2006. KAGRA was approved in 2012, and its distinguishing feature is for the mirrors to be held at low temperatures cryogenically, which will reduce thermal noise. Its two interferometer arms are each 3 km long, and the tunnels to hold them were completed in 2014. The project was troubled by water in the tunnels for a couple of years, but the initial phase of operations, i-KAGRA, at ambient temperature, took place in 2016. A picture of KAGRA taken in 2019 showing the place where the two interferometer arms converge is shown in Fig. 8.15.

The initial cryogenic operation, "b-KAGRA" has gone into operational phase in early 2020. KAGRA should be able to detect gravitational waves from merging neutron stars out to a distance of 240 Mpc with a signal to

Fig. 8.15 A view of the KAGRA gravitational-wave detector, taken close to the place where the two interferometer arms converge. The laser beam is injected into the system from right. This detector is situated underground, in part of the old Kamioka Mine. KAGRA went into operation early in 2020 while this book was being written. Credit: Kamiokande Laboratory collaboration

noise ratio of 10, with a forecast detection rate of 2–3 per year, and it is now joining in the observing campaigns of LIGO and Virgo.

Looking forward the European Space Agency, ESA, has been planning a space-based gravitational wave interferometer for many years, with the project name LISA, (Laser Interferometer Space Antenna). A prototype proof of concept satellite, LISA pathfinder was launched in 2015, and operated on orbit since 2016, in a heliocentric orbit at the Lagrange point L1. It carried a single short arm, and its main goal was to test the noise level in the system, which was shown to be very close to the low level needed for LISA operations. The final interferometer is not planned for implementation and launch before 2030. Its main advantage over ground-based systems is it sensitivity to low frequency gravitational waves, so that it should be able to explore a lower range of binary merger masses. Its overall sensitivity should allow it to detect many sources within the Milky Way. It will also aim to detect mergers between supermassive black holes, objects with "chirp masses" between 1000 and 10^7 solar masses as far away in time as redshift 15. It should be able to detect the spins of the components of mergers close to z = 3, which will help to show whether they have accumulated mass by steady accretion or by more violent mergers. For nearer mergers LISA should be able to detect them sufficiently long before the merger itself to allow conventional telescopes to obtain information, perhaps shedding light on the formation of quasars. LISA should also be used to determine cosmological distances, with the aim of acting as a "standard siren" to determine the Hubble-Lemaître parameter with uncertainty much less than 0.1 km/sec/Mpc.

Now that the way forward has been shown to be worthwhile, there are two major "third generation" projects being explored for future gravitational wave research. One of these is the Einstein Telescope, an equilateral triangular configuration 10 km per side, which is being projected for an underground observatory. The project is being proposed by a European consortium, and will incorporate new technology: a cryogenic system to cool some of the main optics to 10–20 K (-263 to -253 Celsius), new quantum technologies to reduce light fluctuations, and a set of infrastructural and active noise-mitigation measures to reduce environmental perturbations. Funding for initial studies has been requested from the European Union. A site is expected to be chosen in 2024, and if all goes well the instrument would start to observe in 2035. A range of physical problems will be tackled, including physics near black hole horizons, the physics of neutron stars, and the properties of matter at ultra-high densities. The Einstein telescope will be able to detect black hole mergers well back into the "dark ages" of cosmic time, between the epoch when electrons and protons combined to form neutral hydrogen, and the epoch of formation of the first objects producing energy from nuclear reactions. The United States GW community is working on its own major GW project, the Cosmic Explorer telescope, which will have 40 km long arms, at ground level. Those working on the two projects maintain constant communications, and planning includes eventual collaboration between these two telescopes.

8.9   Stop Press

(a) An exciting, if frustrating, aspect of gravitational wave research for an author of a book is that once the technical breakthrough has arrived, and the researchers make continuous small improvements to their system, there is a steady stream of new discoveries with which only the media can keep pace. Even so I want to note two notable detections published in the summer of 2020 based on observations by LIGO and Virgo in the previous year. One of them, labelled GW190521, was due to the fusion of two black holes whose masses sum to over 120 times the mass of the Sun, with a possible mass for the bigger of the two of over 80 solar masses. This range of masses was deemed by the theorists to be unstable for black holes, which therefore should not exist. New ideas include the possible fusion of already fused black holes inside a large dense star cluster. The galaxy where the merger occurred is too distant to be able to make any kind of complementary testing. We are observing an event which took place when the universe had half its present age. The other event, GW190814, is estimated to have been caused by a black hole of 23 solar masses merging with an object of 2.4 solar masses. The peculiarity is that the experts cannot decide whether the second object is the lightest possible black hole, or the heaviest possible neutron star, or some other kind of more exotic end state following a supernova explosion. These two examples give us a glimpse of the opportunities opening up for astrophysicists in the field of the physics of highly compact objects, made possible by detecting gravitational waves.

(b) The results of the campaign between LIGO and Virgo during 2019 have now been published, and summarised in a figure, which shows the distribution of the different types of objects detected. There are now 50 detections in all, of which 37 were made during this campaign. Some of the

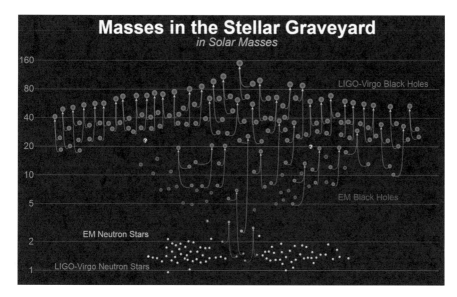

Fig. 8.16 Masses of detected LIGO/Virgo compact binaries. This plot shows the masses of all the merging compact binaries detected by LIGO/Virgo. Black holes in blue, neutron stars in orange. Also shown are stellar mass black holes (purple) and neutron stars (yellow) discovered with electromagnetic observations. Credits: LIGO/Virgo/Northwestern Universityh/Frank Elavsky

detections clearly open new lines of research, because the most massive black holes exceed by half an order of magnitude the highest masses predicted. The population of objects is growing to the point where it will make a major contribution to our understanding of the late stages of stellar evolution (Fig. 8.16).

Further Reading

Books

Book, popular: Brian Clegg, "Gravitational Waves" how Einstein's Spacetime Ripples Reveal the Secrets of the Universe. Icon Books, 2018

Book, general audience: Schilling, Govert. "Ripples in Spacetime, Einstein, Gravitational Waves and the Future of Astronomy" Prologue by Martin Rees. Bellknap Press of Harvard University. 2017

Book: Collins, Harry. "Gravity's Kiss" University of Chicago Press. 2017

Hendry, Martin. "Gravitational Wave Astronomy". (TEDx Glasgow talk) General audience

Weinstein, Alan Recent Results on Gravitational waves from LIGO and Virgo, on YouTube. https://www.youtube.com/watch?v=-i_ARhHfbpg; technical

Quite a technical talk: Barish Barry "Einstein, Black Holes, and Cosmic Chirps" https://www.youtube.com/watch?v=BNlK_nSJDyc

9

Cosmic Rays? Cosmic Particles

9.1 The Discovery of Cosmic Rays

The story of cosmic rays is a classical example of science as an international endeavour. It starts with a simple measuring instrument, the electroscope, invented in the eighteenth century comprising two thin leaves of gold leaf suspended from an insulating rod in a glass jar. If the rod is touched by an object charged with static electricity the leaves spring apart, but gradually come together again after a few minutes, as originally found by the French physicist Charles-Augustin de Coulomb in 1725. In 1835 the British physicist Michael Faraday showed that the tendency of the leaves to fall together occurs under almost all circumstances, even when their supporting rod is perfectly insulated (Fig. 9.1).

In 1898 Marie Curie, born in Poland and living in France, who had discovered the radioactive elements radium and polonium, showed that in the presence of radioactivity the leaves of an electroscope come together very quickly. She concluded that the particles emitted by radioactive elements could be detected with the aid of an electroscope. In 1908 through 1909 the Dutch Jesuit physicist Theodor Wulf took an electroscope which he had made more robust to the Eiffel Tower, to test the idea that the particles come from within the Earth. He measured the difference in the rate of approach of the leaves at the foot of the tower and at the top, expecting to find a slower rate at a higher elevation above the ground, but he did not find any measurable difference. The first strong hint that the particles come not from within the Earth but from above was found by the Italian Domenico Pacini, in 1911. He had tested the discharge rate of his electroscope at a number of sites, at different altitudes

© Springer Nature Switzerland AG 2021
J. E. Beckman, *Multimessenger Astronomy*, Astronomers' Universe,
https://doi.org/10.1007/978-3-030-68372-6_9

Fig. 9.1 A simple electroscope. Two fine gold leaves suspended from an insulating rod. They separate if the rod is touched with static electricity, but gradually come together again if left alone, due to the ionisation of the air in the jar by cosmic rays. This was used to show that the cosmic rays originate outside the Earth, by taking an electroscope up in a balloon and showing that the leaves come together more quickly. Credit: Setreset and Marco Angelucci, CC BY-SA 3.0 <https://creativecommons.org/licenses/by-sa/3.0>, via Wikimedia Commons

on the tops of mountains, over a lake, and over the sea, without finding any measurable differences. He then immersed it under the surface of the sea, at a depth of 3 m and found a consistent 20% reduction in the discharge rate. The decisive step was made by the Austrian physicist Victor Hess. In 1912 he flew with his electroscope in a balloon, and made five flights, reaching a maximum altitude of 5.2 km above the ground. Figure 9.2 is a picture of Hess landing after one of the flights.

Fig. 9.2 Victor Hess lands, on August 7th 1912, with his electroscope after a balloon flight in which he showed that the flux of cosmic rays reaching his instrument increased with altitude. Photograph courtesy of the Victor F. Hess Society and the Archives of the Victor F. Hess Research & Heritage Centre, Pöllau

His result was that the discharge rate was over three times greater at altitude, showing that the source of the particles causing the discharge must come from above. This fundamental work by Hess was recognised with the award of the Nobel Prize, but not until 1936. His results were improved on by the German physicist Werner Kolhörster who also made balloon flights, in 1913 and 1914, reaching a height of 9200 m, and finding an increase in the discharge rate of more than a factor 10 compared with the ground. The original plots from the experiments of Hess and of Kohlhörster are shown in Fig. 9.3.

Kohlhörster also found that the cause of the discharge could not be gamma rays, because he could show that the stopping power of the air for the discharging particles was much less than the known stopping power for gamma rays. It is a wry reflection on history that his final flight took place on the day when the Grand Duke Franz Ferdinand of Austria was assassinated, the day

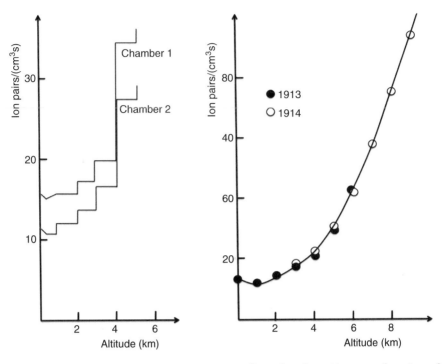

Fig. 9.3 The discharge rate of the electroscope flown by Victor Hess, as a function of height above sea level (left). The same kind of plot, extended by Kohlhörster to higher altitudes (right). These results showed that cosmic rays must be coming from above the atmosphere Credit Alessandro de Angelis via Wikpedia Commons

which marked the effective start of the First World War, and a long pause in the free collaboration of scientists across Europe.

The name "cosmic rays" was invented in 1926 by the United States physicist Robert Millikan, who had been awarded the Nobel Prize in Physics in 1923 for measuring the charge of the electron. He had not at first believed in their origin outside the atmosphere, but after a series of experiments he became convinced, and coined the term, which has been used ever since. We will see that this was in many ways unfortunate, but it was not the first or last time in science when a term is adopted and then persists in spite of being found inadequate. It would be more appropriate to use the term for cosmic gamma rays, which are the highest energy component of light, that is of electromagnetic radiation, and whose photons have energies comparable to those of the cosmic ray particles. The detection techniques for cosmic ray particles and for the highest energy gamma rays have much in common, so we will cover both of them in this chapter.

9.2   What Are Cosmic Rays?

Going straight to the point cosmic rays are not rays, they are particles. We normally use the term rays for light, and any other form of electromagnetic radiation, all the way from the radio range through the infrared, the visible, the ultraviolet, X-rays, and gamma-rays, all of which have been dealt with in their respective chapters of this book. These all travel at the speed of light. A particle is the generic term for any small object which has mass, technically described as an object with rest mass, and which cannot ever reach the speed of light, although the particles we will meet in the present chapter and in the chapter on neutrinos travel at speeds very close to light speed.

9.2.1   Cosmic Ray Composition, and the Abundances of the Elements

The majority of the cosmic ray particles are protons, the nuclei of hydrogen atoms, which are the simplest and most abundant types of atoms in the universe. In second place come the helium nuclei also known as alpha particles, from the days of their discovery in terrestrial radioactivity at the end of the nineteenth century; they are made up of two protons and two neutrons. Then, in much smaller numbers, come the group made up of lithium beryllium and boron, followed by a set of heavier elements, carbon, nitrogen, oxygen, sulphur, and others. In addition there are also cosmic ray electrons. In the Fig. 9.4 you can see the relative numbers of the cosmic ray nuclei, plotted in order of their atomic numbers, which give the number of protons in each of their nuclei, so the scale is one of increasing atomic number. These data were taken from rocket and satellite experiments towards the end of the twentieth century, but the measured values of the abundances have changed little in more recent measurements.

The most complete modern experiment which gives us the composition of cosmic rays is the Alpha Magnetic Spectrometer (AMS) shown in Fig. 9.5 which was built by a team led by the Nobel Laureate Samuel Ting, and was placed on the International Space Station in 2011 using the last of the Space Shuttle flights.

This spectrometer is aimed at a number of different observational experiments on cosmic rays, including a key search for antiparticles. One interesting result has been the detection of a significant fraction of positrons, in the range up to 20% of the electrons, a fraction which peaks at 275 GeV and comes uniformly from around the sky. These could be produced by the annihilation

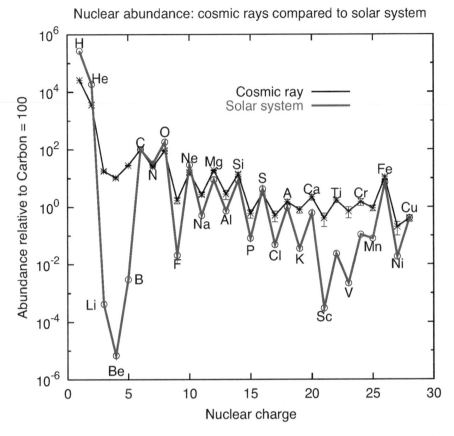

Fig. 9.4 The abundances of the chemical elements, by atomic number, observed in cosmic rays from rockets and satellites outside the atmosphere compared with the abundances observed in the Sun and stars. Three of the lightest elements, lithium, beryllium and boron are far more strongly represented in the cosmic rays than in the other sources. Credit: Juan Aguilar

of dark matter particles in space, which would be an exciting result, but other causes cannot be excluded. More generally the experiment looks for antihelium nuclei (antialpha particles) composed of antiprotons and antineutrons, as evidence for antimatter as a fundamental constituent of the universe. Until now only an upper limit of one part antihelium to a million parts of helium has been established.

In Fig. 9.6 you can see the numbers of counts of a large range of atomic nuclei which have been detected with the AMS, which is an excellent tool for measuring their relative abundances. The ratios of the elements found in an astronomical context are called the "abundances" of the elements. They are usually expressed with reference to hydrogen, the most abundant element.

Fig. 9.5 The Alpha Magnetic Spectrometer (AMS) on board the International Space Station, on an exterior surface, photographed by an astronaut during a space walk Credits: NASA/AMS Collaboration/CERN

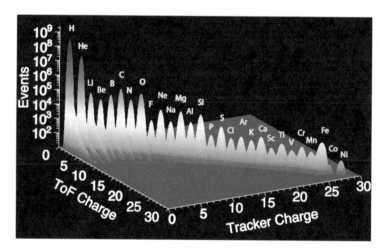

Fig. 9.6 The number of counts of atomic nuclei falling onto the collector plate of the AMS in a specific time interval. The proportions are comparable to the element abundances measured in stars except for the elements lithium (Li), Beryllium (Be), and Boron (B), which are mostly produced in interstellar space. The numbers of events are a combined representation of the counts from two separate types of detectors. Credits: NASA/AMS Collaboration/CERN

Starting early in the twentieth century astronomers have used spectroscopy to measure element abundances in stars and in the interstellar medium, and since the 1950s they have been able to match these measurements to the theory of element production. To give a very broad summary, hydrogen, and over 80% of the helium in the universe were produced in the Big Bang, along with deuterium, and some 10% of the lithium. The remaining nuclei have been "cooked" within stars. The lighter group of elements, including oxygen, carbon, nitrogen, and sulphur, have been produced continually during the lifetimes of stars with moderate mass, and expelled into space either via stellar winds, or by relatively mild explosions, the novae. Heavier elements, including iron and nickel, were produced within more massive stars, or within specific phases of the interchange of material between binary stars, during their violent explosive phases, the supernovae. The systematics of many of the physical processes involved were presented in a classical paper in 1957 by the British astronomers Margaret and Geoffrey Burbidge, and Fred Hoyle, with the US astronomer William Fowler (who was awarded the Nobel prize in 1983 for this and related laboratory work). This paper, often referred to by the nickname B^2FH went a long way to setting the basis for all subsequent studies of the origin of the elements and of their abundances, by attributing them to specific nucleosynthesis processes within stars. But the abundances of the light elements lithium, beryllium, and boron, (collectively LiBeB) as measured in the spectra of stars are orders of magnitude greater than the values predicted by any of these processes. This is essentially because these elements, as well as deuterium (heavy hydrogen, D) are known to be destroyed in the conditions within stellar cores, even when they do take part in the synthetic pathways predicted by the theorists. But cosmic rays came to the rescue. In 1971 the French physicists Maurice Meneguzzi, Jean Audouze, and Hubert Reeves showed that high energy particles in interstellar space can produce significant quantities of LiBeB, by splitting the nuclei of carbon, nitrogen, and oxygen, whose nuclei are larger. This process is called spallation, and it was possible to estimate the cosmic abundances of the LiBeB produced by interstellar spallation due to cosmic rays. The results for Be and B, in particular, showed that cosmic ray spallation could produce all the observed amounts of these two elements. This was not so for Li, and the sources of cosmic Li abundance have been the subject of a great deal of research, still without a definitive result, which has implications for Big Bang cosmology. We can see from Fig. 9.4 that the abundances of LiBeB in cosmic rays are orders of magnitude higher than those in stars, which does indicate their almost certain origins in interstellar spallation processes. To understand the abundances of the chemical elements we need to bring together the physics of element production, and the chemistry of the combinations of elements in space, in the stars, and in the planets, which we need to understand in order to obtain reliable measurements to test the physics. In addition we also need to know the chemistry of meteorites, and nowadays of comets, geology and planetology. This is multi-messenger astronomy at its most relevant, and it was relevant well before the measurements of gravitational waves and of cosmic neutrinos.

9.3 When Cosmic Rays Played a Key Role in Particle Physics

9.3.1 The Discovery of the Positron

Even before the astrophysical origins of cosmic rays could be deeply investigated they began to play an important role in the new science of particle physics. In 1928 the British physicist Paul Dirac had proposed the existence of a particle with the properties of the electron, except that it should have a positive rather than a negative electrical charge. In 1932 the United States physicist Carl Anderson published the first paper on the particle, which he had detected using cosmic rays. The editor of the journal the Physical Review, suggested the name positron, and Anderson agreed. In fact several physicists, including a Russian a Chinese-American, and a French couple, had detected positrons in their experiments previously but had not realised the importance of their detections. The way in which Anderson made his discovery is shown in Fig. 9.7 which is one of 15 photographs in which Anderson found evidence for positrons, out of a total of 3000 analysed. The photograph shows a cloud chamber, in which a charged particle produces a track of tiny droplets of condensed water, which can then be photographed. The chamber has a magnetic field applied which makes the particle take a curved path. In the middle of the chamber is a lead plate, which slows down the particle. The track of a slower particle has greater curvature, and this allows the observer to show which way the particle travelled. In this case upwards in the photograph. Knowing the direction of travel of the particle and the direction of the curved path gives the sign of the charge on the particle, and in this case it was a positive charge. The length of the track shows what kind of particle it was. In this case the length was similar to those of electron tracks, and not those of proton tracks, which would be much shorter as a proton loses energy quickly by collisions with the atoms in the chamber. So Anderson could conclude that the particle was a

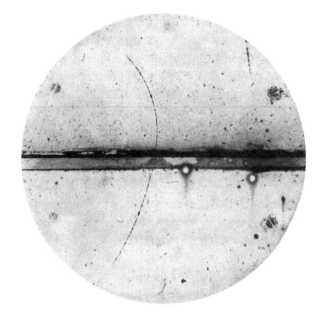

Fig. 9.7 A photograph of one of the first positrons discovered, by Anderson in 1932, using cosmic rays. The particle detected in a cloud chamber passes upwards through a lead plate in the middle, which slows it down. This shows the direction of flight, from bottom to top. The curvature of the track, in an imposed magnetic field, shows the sign of the electric charge of the particle, positive, and the length of the track shows it to be an electron, in this case a positive electron, i.e. a positron. Credit Carl Anderson,1933, Physical Review Vol. 43, p. 491

positive electron, that is a positron. Anderson received the Nobel prize for his discovery, along with Hess, in 1936, although he clearly had much less time to wait!

9.3.2   Cosmic Rays Lead to the Discovery of a Number of Subatomic Particles

In the period between the First and the Second World Wars, a period when physics began to advance at an ever increasing pace, a period when much of the foundation of modern physics, above all the theories of quantum mechanics and relativity, were laid down, one of the main tools of post-Second World War physics, the high energy particle accelerators, had not yet been well developed. It is true that in 1932 John Cockroft and Ernest Walton had built a high voltage machine which could accelerate particles through a potential of almost one million volts, and had split the lithium nucleus, (for which they were awarded the Nobel Prize in 1951), and also in 1932 Ernest Lawrence had developed the first cyclotron, (for which he was awarded the Nobel Prize in 1939) with which he accelerated particles to 1.25 million volts, and split a nucleus just weeks after Cockroft and Walton. But these energies were very small compared to those of cosmic ray particles, and physicists understood this well. So from the 1930s to the 1950s the main line of attack for those wanting to break new ground in elementary particle physics was to use cosmic rays to collide with selected targets of local nuclei and examine the particles which were produced. In 1936 Carl Anderson and Seth Neddermeyer used a platinum plate to intercept cosmic ray showers, and to detect their products in a cloud chamber. One of the particles they detected with consistency had 200 times the mass of an electron, but the same single positive charge. In 1934 the Japanese physicist Hideki Yukawa had predicted the existence of a particle with a mass in this range, and with the same charge. This had become known as a mesotron (later elided to "meson") because its mass had to be intermediate between that of an electron and a proton. It was first thought that the particle discovered by Anderson and Neddermeyer was Yukawa's meson, but later it was shown to have rather different properties. We now know this particle as a mu meson, normally shortened to muon. The particle predicted by Yukawa was in fact discovered in 1947 in another experiment with cosmic rays, by the Brazilian Cesar Lattes, the Italian Beppo Occhialini, and the Briton Cecil Powell. The latter was awarded the Nobel Prize in Physics in 1950 for his development of the photographic method for detecting elementary particles in cosmic ray showers. This particle was named the pi meson, shortened to pion. During the late 1940s and early 1950s a number of new subatomic particles were discovered using cosmic ray interactions. These included the k meson, or Kaon, discovered by the US physicists George Rochester and Clifford Butler in 1947, and the Λ hyperon discovered in Australia by Victor Hopper and Sukumar Biswas in 1950. One of the reasons why cosmic rays played such a major role in particle physics in that period is that the experiments involved were inexpensive, and the world was still in a phase of economic recovery from the major setback of the Second World War.

9.4   The Energy Distribution of Cosmic Rays

In Fig. 9.8 we show the number of cosmic ray particles which impinge on the Earth in a given time as a function of their energies. Now, with the AMS in orbit, we can obtain these figures directly, without needing to correct for the absorption of the particles by the atmosphere.

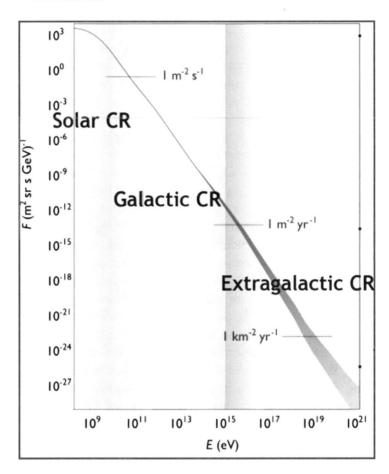

Fig. 9.8 The number of cosmic rays which fall onto the Earth in a given time, as a function of their energies. The horizontal axis gives the energy in electron volts, and the vertical scale is the flux in particles per square metre per steradian, per second, per GeV. Credit: Sven Lafebre CC BY-SA 3.0 <http://creativecommons.org/licenses/by-sa/3.0/>, via Wikimedia Commons

To explain the scales on the axes, the x-axis (horizontal) gives the scale of the energy of a particle in electron volts (eV), and the y-axis gives the scale of the flux of particles at each energy. These scales are worth a brief explanation. An electron volt is the energy acquired by an electron when it is accelerated by the potential difference of one volt in an electric field. This is a practical unit, because particles are often accelerated using electric fields, albeit usually with ranges of megavolts or greater. Electrons are tiny particles and an electron volt is quite a small amount of energy, in fact 1.6×10^{-19} joules (to give a basic figure, there are 4.2 joules to one calorie, and a calorie is the energy needed to heat up one gram of water by one degree Celsius, so the number of joules in an electron volt is very tiny indeed). Particle energies in accelerators and in cosmic rays are usually found in the range of Megaelectron volts (1 MeV = one million electron volts) or Gigaelectron volts (1 GeV = a thousand million electron volts) or higher. The electron volt is a standard energy unit, which is used for all particles, not just electrons. The units of the number flux are in particles per square metre per second per steradian per GeV. This reflects the fact that cosmic rays come from all over the sky, and not in a parallel beam onto the Earth; so we need to measure not only how many particles pass through a square metre in a second but also how many come from a given fraction of the sky. A steradian is a unit of solid angle, and the full sky, visible down to the horizon from any point takes up 2π steradians. We can see in the figure that the higher the energies of the particles the fewer there are in the constant rain of cosmic rays hitting the Earth. For energies around 10^9 eV, which is the same as 1 GeV, about 10 particles per square metre are incident on the Earth every second. The numbers fall off very quickly as we go to higher energies. When we reach the highest energy to which particles have been accelerated in the giant Large Hadron Collider at CERN in Geneva, which at the time of writing is $1.4 \times 2 \ 10^{13}$ eV (14 TeV), the number is around 1 particle per square metre every 10 days. But we can see that cosmic ray experiments have been able to detect particles of much higher energy, reaching up to 4×10^{20} eV, 30 million times more energetic than the particles at CERN. A single particle with this energy could heat up a gram of water by 30 degrees Celsius, and it could clearly inflict real damage on a human body if it happed to strike someone. Luckily there are very few of these particles around, of order one per ten thousand square kilometres per year.

9.5   Where Do Cosmic Rays Come From?

As we see in Fig. 9.8 the great majority of cosmic rays have energies less than 10^8 eV, (conventionally expressed as 100 MeV). It is relatively straightforward to show that most of these are particles produced by the Sun, and blown out by the solar wind in a stream which sweeps past the Earth towards the limits of the solar system. It is not hard to show this because of the general direction from which they come. Even so they do not follow straight line trajectories radially outwards from the solar surface, but their paths are deviated somewhat by the outer solar magnetic field. In order to identify the source of a particularly strong burst of the particles, from a solar storm, we need the help of imaging which shows us the chromospheric and coronal features which produce these particle bursts. This problem of identifying the origins of cosmic rays is not acute for the solar component, but the fact that over long trajectories through space the paths of cosmic ray particles are always subject to small scale magnetic fields at the point of origin, and large scale magnetic fields between the origin and the Earth means that it is rarely possible to identify directly a source of high energy cosmic rays from within the Galaxy or from an extragalactic object. This problem can be eased if a cosmic ray source also emits gamma rays or neutrinos, because both of these components travel to us along tracks which are not deviated by magnetic fields. This point promises to be important although very few cases of such direct detections of cosmic ray sources have been possible until now. Solar wind particles cross the Earth's orbit at rates in the range one particle per square metre per second.

The cosmic rays of intermediate energies arrive at the Earth from most directions on the sky. The chief candidates for the production of these particles are the supernovae explosions which mark the deaths of stars with relatively high masses.

The highest energy cosmic rays are produced in the processes which release particles of the highest energies in the universe. They are the active galactic nuclei, but also the mergers between black holes and neutron stars, of which we are just beginning to learn through the new messenger of gravitational waves.

9.6   Cosmic Ray Telescopes and Observatories

9.6.1   Ground Based Telescopes

When we describe ground based cosmic ray telescopes it is necessary to understand that we need to widen the traditional definition of cosmic rays as consisting of high energy particles to include the highest energy photons: the gamma-rays. This is because the techniques for detecting cosmic rays on the ground also detect gamma rays, and because rays and particles of such high energy are in any case produced in similar types of astronomical sources. We saw how the AMS experiment on the International Space Station, outside the atmosphere, gives us the best statistics of the different species of atomic nuclei in the cosmic ray flux. But just as optical telescopes in space cannot generally have very large collecting areas so that ground based telescopes are needed to study very distant faint sources in detail, the fluxes of the highest energy cosmic particles and gamma rays are low, and large collecting areas require ground based observatories. There are a number of techniques used to for comic ray and gamma ray detection on the ground, which use the Earth's atmosphere as part of the detecting system.

Figure 9.9 shows typical results of the collisions of a gamma ray, and also a high energy cosmic ray particle, with nuclei in the upper reaches of the atmosphere. In both cases the result is a shower of particles, but the compositions are different. The high energy gamma ray produces an electron-positron pair, but when the positron meets an electron in the atmosphere the two particles

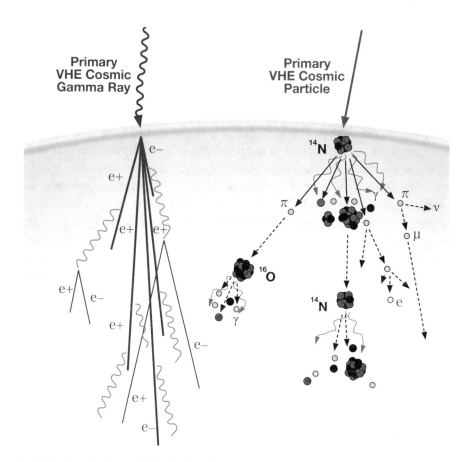

Fig. 9.9 Descriptions of the kinds of reactions which occur when very high energy (VHE) particles and very high energy gamma rays strike the upper part of the atmosphere. The diagrams represent the collision of a single gamma-ray (left) and a single cosmic ray particle (right) with the nucleus of an atom. The gamma-ray gives rise to a shower of high energy electrons. The cosmic ray particle has a more complex effect, producing a shower including muons, pions, electrons, and neutrinos. The two types of showers can be distinguished by the light patterns they produce in the detectors on the ground. Credit Konrad Bernlöhr

annihilate and produce two gamma rays, which in turn can collide with nuclei and give further electron-positron pairs. This is converting photons into particles with mass, back to photons again, and back to particles with mass, which can take place only in the presence of other particles, in this case the nuclei of the atoms in the upper atmosphere. This process continues until the energy of the initial gamma-ray has been divided into a whole shower of electrons, many of which are moving at very high velocities. When a high energy cosmic particle (a conventional cosmic ray) hits the upper atmosphere and collides with the nucleus of an atmospheric atom, nitrogen or oxygen, it produces a variety of different particles, including neutrinos, pions, electrons, and also gamma rays. All of these particles are moving at very high speeds.

When a particle moves at very high speed through a medium an effect is produced which is analogous to the sound shock wave produced when a body moves through a medium at supersonic speed, the famous "supersonic bang". This "superluminal bang" is produced when particles move through a medium at velocities greater than that of light. It is important to note that this phenomenon does not violate one of the principles of special relativity, that nothing can move faster than light, because the principle really states that nothing can move faster than light in a vacuum. Light in any medium moves a tiny fraction more slowly than light in a vacuum, but this tiny fraction is sufficient for many of the particles in the showers produced by high energy cosmic rays and gamma rays to be indeed moving faster than light in the atmosphere through which they pass. This leads to the "superluminal bang"; from this point on we will refer to this by its accepted name in physics, the Cherenkov effect, called after the Russian physicist Pavel Cherenkov, who first observed, and interpreted it, and was awarded the Nobel Prize in 1958. The Cherenkov light from atmospheric particle showers is now used as a key tool to detect and measure high energy cosmic rays impinging on the upper atmosphere. Detectors using the Cherenkov light from cosmic ray showers prove to be the most effective way of detecting the showers and of inferring the nature of the high energy particle, or the gamma rays which produced them.

9.6.2   The Pierre Auger Observatory

The biggest cosmic ray observing instrument in the world (and indeed physically the largest astronomical instrument in the world) is the Pierre Auger Cosmic Ray Observatory. Its purpose is to detect cosmic rays with energies greater than 10^{17} eV, and try to find their origin. It uses two different ways of detecting cosmic ray particles via their air showers: a set of Cherenkov detectors, and a set of telescopes to detect fluorescent emission from the ionised particles in the shower. The Cherenkov detectors comprise 1660 water filled tanks, one of which is shown in Fig. 9.10a, spread over an area of 3000 square kilometres, as shown the plan in Fig. 9.10b. Combining the information from the two sets of "telescopes" it is possible to distinguish whether the shower of particles detected was initiated by a gamma-ray or a cosmic ray particle.

Each detector tank is filled with pure water in a polyethylene container, with photomultiplier detectors in their upper parts. A high energy particle produces Cherenkov radiation within a detector, which is detected by normal photodetectors sensitive to visible light. Combining the timing and strengths of the signals in the array it is possible to estimate the energy of the cosmic ray, and also the direction from which it has hit the atmosphere. The observatory also has four fluorescence detectors which make direct measurements of the amount of "normal" (non-Cherenkov) light emitted from the ionised gas in the atmosphere due to a particle shower. These are complementary to the Cherenkov detectors; they allow calibration of the total energy of the particles detected by the surface detectors in the array.

The highest energy cosmic rays are few and far between, they arrive at rates of order one per square kilometre per year but they have an advantage in that their trajectories are less affected by the magnetic field of the galaxy. Each such cosmic ray has an energy over a million times greater than that of a proton in the large hadron collider at CERN, the highest energy particle accelerator in the world. In 2017 the Pierre Auger team from 18 countries declared that they had accumulated detections of these extremely high energy cosmic rays enough to show that they must come from outside our Galaxy, even though the exact point of origin cannot be well defined. The measurement showed that for the 30,000 particles detected with energies of 2 joules (1.2×10^{19}eV) there was an imbalance of 6% between hemispheres, with a peak axis 120° away from the centre of the Galaxy where there is a tendency for a high concentration of external galaxies The Auger observatory is waiting constantly for even rarer events of even higher energy, which will allow the direction of origin to be pinned down with greater accuracy.

a

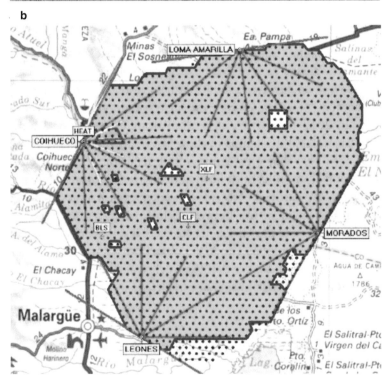

b

Fig. 9.10 (a) One of the 1600 water tank detectors of the Cherenkov radiation produced by the high energy particles in a shower produced by an ultra-high energy cosmic ray. These form part of the Pierre Auger cosmic ray telescope, situated in Argentina. (b) A map of the surface detectors, which use the Cherenkov effect to detect and

9.7   Steerable Telescopes for High Energy Gamma-Rays

I am incorporating this section in the chapter on cosmic rays because the gamma rays observed have energies high enough to be comparable to those of cosmic rays, and their sources are strongly related to the sources of cosmic rays. And as you know from earlier in this chapter, they are in a strict sense the only cosmic "rays". The steerable telescopes are designed to detect gamma rays in the range from 30 GeV upwards. The limit at the high frequency high energy end of their range is not instrumental but is due to the fact that gamma-rays at TeV energies are strongly absorbed when they interact with the general extragactic background light and do not reach us. Even so, powerful sources in this range could be detected if they are far enough away so that their emitted frequency is reduced by cosmological redshift. The steerable gamma-ray telescopes are large, much bigger than orbiting gamma-ray telescopes, which is their principal advantage. They take advantage of the interaction of the gamma-rays with the atmosphere, and their direct detection is by the Cherenkov light accompanying the relativistic particle shower produced by the original incoming gamma-ray. There are now a number of these large telescopes; they include VERITAS, HAWC, and HESS. The latter is an acronym for High Energy Spectroscopic System, an acronym carefully chosen to honour Victor Hess, whose balloon-borne measurements we have described early in the chapter. It comprises five telescopes, four of 12 m diameter and a central major telescope 28 m in diameter, and detects gamma-rays from 30 GeV to 100 TeV energy.

9.7.1   The MAGIC Telescope

But here I will give a more detailed account of MAGIC, the two-telescope gamma-ray observatory because it is at my home institution so I am directly familiar with it. One of its two telescopes is shown in Fig. 9.11.

The acronym means Major Atmospheric Gamma Imaging Cherenkov. The telescopes are designed to point to a specific known astronomical source and to detect the Cherenkov light images from gamma rays in a similar energy

---

Fig. 9.10 (continued) measure the showers from ultra-high energy cosmic rays. They cover an area of 3000 square kilometres. Each circular dot corresponds to a detector. The four fluorescence detectors, at the foci of telescopes, are at the points of convergence of the brown straight lines. Credits: Pierre Auger Observatory Collaboration

Fig. 9.11 One of the two MAGIC gamma-ray telescopes at the Roque de los Muchachos Observatory, La Palma, Canary Islands. The mirror diameter is 17 m and the Cherenkov radiation detector is on the parabolic shaped boom at the prime focus of the telescope. Credit: The Magic Collaboration

range to that described for HESS. The main challenge is to separate the electromagnetic showers produced by gamma-rays from the generally isotropic background due to charged particle (mostly hadron) showers. This is done automatically by specially designed software trained on the different morphological patterns of light from the two types of events, and is extremely important, since the hadron-induced showers may produce much more Cherenkov light than the gamma-ray showers. MAGIC I has been operational since 2004, and MAGIC II joined it in 2009. The two telescopes, separated by 85 m on the ground, are operated together in "stereoscopic" mode, such that only those events detected by both telescopes within a brief time window of 180 nanoseconds are recorded and analysed.

One of the prime earlier discoveries with MAGIC was the discovery in 2008 that the Crab Nebula pulsar is emitting gamma-rays in the GeV range. This was followed in 2011 and 2012 by extension of the energy range up to 400 GeV, and in 2016 to 1.5 TeV the highest energy pulses found in any object. Also MAGIC detected constant low level gamma-ray emission from the Crab in the intervals between the pulses. These findings were not predicted by the theoretical models of pulsar emission, and are still difficult to explain. The Crab Nebula pulsar has been a pathfinder object since it was first detected at radio frequencies. We have seen how it was the first pulsar to be detected at optical wavelengths and now also in gamma-rays. MAGIC is finding enough data to contribute to new models to explain how the pulsar emission varies with energy. Another exciting measurement with MAGIC was the observation of a gravitational lens at very high energies. In July 2014 the gamma-ray telescope Fermi-LAT on board the Fermi satellite detected a burst of gamma-rays from a supermassive black hole in a quasar QSO B0218 + 357. This is at 7 GLy (seven Gigalight years) from us, and the radiation is refocused by the gravitational field of a cluster of galaxies 6 GLy away making it bright enough to detect. FermiLAT scans the whole sky every 3 h which makes it ideal for detecting outbursts, but for deeper observations ground based telescopes were alerted. Unfortunately the night was moonlit, so the Cherenkov telescopes on the ground could not be used. However it was known that this gravitational lens gives two paths for the gamma-rays, and the second path takes 11 days longer, as had been shown at radio wavelength during a previous outburst from B0218 + 357, in 2012. By that time in a moonlight-free night MAGIC made the detection predicted, at much higher energies than those which could detected by FermiLAT. The delay of 11 days, timed exactly, coincided with the delay previously measured for the radio burst, and this is one of the growing set of corroborations of the theory of General Relativity. A diagram is given in Fig. 9.12.

Another exciting multi-messenger event in which MAGIC played a key role was triggered by the measurement of a 290 TeV neutrino in the Ice Cube

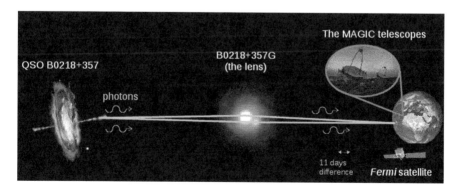

Fig. 9.12 Gravitationally lensed gamma-rays from a distant quasar detected first by the Fermi satellite, then also at much higher energy by the MAGIC telescopes 11 days later Credits: NASA/ESA; AGN and Fermi images credits: NASA E/PO—Sonoma State University, Aurore Simonnet; visual editing: Julian Sitarek, the MAGIC Collaboration

neutrino observatory in Antarctica. Its position was well enough determined to associate it with a kind of quasar called a BL Lac object, with identification TXS 0506 + 56, which showed enhanced gamma-ray activity as measured by FermiLAT. The source was monitored by MAGIC for 41 h and gamma-ray variability was well detected. With data from the three instruments models could then be built, showing that the Very High Energy gamma-rays could be produced by the interaction of hadrons (essentially protons) with photons, while the lower energy gamma-rays corresponding to the Fermi detection were most likely due to interaction of leptons (essentially electrons) with photons, a process called inverse Compton scattering. All of the processes must be occurring in an extremely high energy jet coming from the quasar nucleus, containing protons with energies up to 10^{18} eV.

A further example of the intervention by MAGIC in a very relevant discovery was the detection and analysis of a gamma-ray burst (GRB) in 2019. In the chapter on gamma-ray astronomy we saw that a gamma-ray burst is a very brief surge, a flash of gamma rays, which must be due to the collapse of a massive star, or the merging of neutron stars in some distant galaxy. The general rate of GRB's is about one per day, and they last from a fraction of a second to hundreds of seconds, followed by a period of less bright but longer lasting emission over a wide range of wavelengths, the afterglow. The GRB in question was detected independently by two satellite observatories, the Neil Gehrels Swift observatory and the FermLAT instrument. The burst was termed GRB 190114C. Within 22 seconds its coordinates were disseminated as an electronic alert to astronomers world-wide. MAGIC has an automatic system which responds to these alerts by pointing the telescopes which are designed to be light so that they can slew rapidly to any point in the sky. In this case within 25 seconds, 50 seconds after the initial detection, MAGIC was observing the GRB. Analysis of the data from MAGIC showed that it had detected TeV photons, and was the brightest source of TeV gamma rays known up to that time. An astronomical telegram on January 15th 2019, after rapid verification of the data, claimed this new observation. The afterglow was followed up by tens of groups observing over a wide wavelength range, providing a comprehensive description of the phenomenon, for the first time, from TeV to radio energies. One optical result was a measurement of redshift, the recession velocity of the galaxy in which the GRB took place, putting it at a distance of 4.5 GLy (Giga light-years). The presence of TeV gamma-rays in GRB's, which has been supported by similar measurements with the HESS telescope, shows that the cosmic explosions which cause them do not only produce synchrotron radiation, enough to explain the GeV gamma-rays, but also inverse Compton radiation, and are even more powerful than originally thought.

9.7.2   The Cherenkov Telescope Array: Future Prospects

The field of gamma-ray astronomy is one of the new fields which is linked to the high energy processes producing cosmic rays, and neutrinos. It is in the course of obtaining a major new boost in the form of the CTA, the Cherenkov Telescope Array, a project typical in its organisation to many of the Big Science projects which are leading the way in experimental studies. It is organised by a consortium of 200 institutions in 31 countries, with over 1500 scientists involved. The CTA project is constructing two observatories, CTA-north, on the island of La Palma, and CTA-south at Cerro Paranal, Chile, both sites with existing major optical and infrared observatories. The heart of the project will comprise 40 medium sized telescopes, with 12 m primary mirrors, aimed at the energy range between 100 GeV and 10 TeV; 16 of these will be at the CTA-north site, and 24 at the CTA-south site. There will be 8 large sized telescopes, with 23 m primaries, 4 in the north and 4 in the south, to extend the range below 100 GeV. The third element of the observatory will comprise 70 small sized telescopes, each 4 m in diameter, to observe the highest energy range, above 10 TeV. It is assumed that most of these gamma-rays will come from within our Galaxy, because those originating in more distant sources will be too attenuated to detect.

As the centre of the Galaxy is overhead at CTA-south, all the small sized telescopes will be sited there. Figure 9.13a and b show CTA-north and CTA-south,respectively, as they will look when completed, and Fig. 9.14 shows the prototype large-size telescope at the Roque de los Muchachos Observatory in La Palma.

Although it is large, its very light structure allows it to point to any source on the sky within 20 seconds, to react to bursts of gamma-rays detected by survey instruments, notably the Fermi satellite gamma-ray observatory. Prototypes of all three sizes of telesccopes have been built and are operational. The full CTA should be working in the early 2020s.

a

b

Fig. 9.13 (**a**) (Upper) The site of the proposed CTA-north observatory, Roque de los Muchachos, La Palma, Canary Islands, with an artist's impression of the telescopes in place. (**b**) (Lower) The site of the proposed CTA-south observatory, Cerro Paranal, Chile, with artist's impression of the telescopes in place. Credits: CTA project/IAC/Gabriel Pérez

Fig. 9.14 Prototype Large Size Telescope of the CTA north, at Roque de los Muchachos Observatory, La Palma. This is a 23 m diameter Cherenkov telescope, now in place and working Credits: CTA project/IAC

Further Reading

Book dedicated mainly to cosmic rays: Probes of Multimessenger Astrophysics. Maurizio Spurio. Springer ISBN 978-3-319-96854-4. Chapters: An overview of Multimessenger Astrophysics. Charged Cosmic Rays in our Galaxy. Direct Cosmir Ray Detection: Protons, Nuclei, Electrons and Antimatter. Indirect Cosmic Ray detection: Particle Showers in the Atmosphere. Diffusion of Cosmic Rays in the Galaxy

Wikipedia article, quite short but covers useful ground at a good level. https://en.wikipedia.org/wiki/Cosmic_ray

Introductory book chapter on high energy cosmic rays. Zbigniew Szadkowski (August 22nd 2018). Introductory Chapter: Ultrahigh-Energy Cosmic Rays, Cosmic Rays, Zbigniew Szadkowski, IntechOpen, DOI: https://doi.org/10.5772/intechopen.79535. Available from: https://www.intechopen.com/books/cosmic-rays/introductory-chapter-ultrahigh-energy-cosmic-rays

10

Cosmology and Particle Physics

10.1   Nucleosynthesis in the Big Bang

Present day cosmology started with the discovery of the microwave cosmic background radiation (CMB) in 1964, by Penzias and Wilson, described briefly in Chap. 2. This swung the balance definitively between the Steady State theory of cosmology and the Big Bang theory in favour of the latter. In the Steady State theory the universal expansion of the galaxies on a large scale, fully established by observations in the four decades since its discovery by Edwin Hubble in the 1920s, was explained as due to the continuous outward pressure of a creation field. In this field new particles are created at a uniform rate everywhere in the universe, and as time progresses they condense into the stars and galaxies we observe. On this theory the universe, seen globally, has a constant mean density of matter, invariant with time. In the 1950s the theory had already received a major blow when the radioastronomers under Martin Ryle showed that as they looked further away, and hence further back in time, the average density of radio sources was bigger, considerably bigger, than it is today. The best interpretation of this was that the universe is not in a steady state, but evolving, from a previously denser state. So the scenario of an origin in a very dense, very hot state, previously proposed by the Russian-American physicist George Gamow, began to be favoured. One of the originators of the Steady State theory was Cambridge astrophysicist Fred Hoyle, and it was he who invented the name "Big Bang" for the rival evolutionary model, in a jokey slightly depreciative manner. But Hoyle was not only a specialist in cosmology. His interests were wide, shared with a bright generation of post-Second World War astronomers in Britain, and included theoretical

© Springer Nature Switzerland AG 2021
J. E. Beckman, *Multimessenger Astronomy*, Astronomers' Universe,
https://doi.org/10.1007/978-3-030-68372-6_10

explanations for the behaviour of comets, and above all really impressive work on the origin of the chemical elements. The seminal article describing in considerable detail the processes of the production of the chemical elements in stars was written by a foursome of astronomers, Margaret and Jeffrey Burbidge, Fred Hoyle, and American nuclear physicist Willy Fowler. This article set on a firm footing a large volume of subsequent astrophysical research involving the elements, their production and distribution, and the use of the elements as tracers of the history of evolution of stars and galaxies. It showed that almost all of the elements heavier than helium are produced inside stars in a series of nuclear processes, some of which take place in a steady way during the lifetime of a star, others taking place only in the extreme conditions just prior to a supernova explosion. Fowler, a nuclear experimental physicist, was awarded the Nobel Prize for this work, and the fact that Hoyle was not included in the award has remained controversial.

It was recognised generally that the starting point of stellar nucleosynthesis had to be hydrogen, which comprises 74% of the mass of the elements in the universe as a whole. What was less clear was how the abundance of helium came to be some 24% by mass. Given the time available for the evolution of the stellar populations in galaxies, and the known physics of the conversion of hydrogen to helium in the conditions in the centres of stars the maximum amount of helium which could have been produced in the universe by stars is considerably less than 10% by mass. Fowler and Hoyle were well aware of this, and so as an exercise they set out to calculate how much helium could have been produced in the extreme high temperatures and densities within the "primordial fireball" of a big bang model universe. It says much for their scientific curiosity that they did this while still being convinced that the primordial fireball had not really existed. The article with their work was authored by Robert Wagoner, a graduate student of Fowler, together with Fowler and Hoyle. Their models including a dense primordial fireball were able to produce the mass fraction of around 25% for the helium observed in the stars, but to satisfy their underlying scepticism they also showed that very massive stars with very high core temperatures might be able to synthesise the helium. In fact their work was following a line initiated in a famous article by Alpher, Bethe, and Gamow in 1948, who had tried to model the synthesis of all the elements in a primordial fireball, but had failed to obtain the observed values for the heavy elements. Although I have used the term "helium abundance" I should really have been talking about ^{4}He, a helium nucleus made up of two protons and two neutrons. The chemical behaviour of an atom, and hence its name, depends on the number of protons in its nucleus. Normal hydrogen, the lightest element, has one proton, but heavy hydrogen or deuterium has a proton plus a neutron in the nucleus, and a heavier, even rarer isotope, tritium, has two neutrons. The next lightest element, helium has two protons and it comes in two forms, the more stable form, ^{4}He, with two protons plus two neutrons and a less stable form, ^{3}He, with two protons but only one neutron. In the primordial fireball ^{4}He was produced by the fusion of four hydrogen nuclei, and ^{3}He was also produced, by the fusion of three hydrogen nuclei, but in a much smaller proportion as it is much less stable.

The details of the nuclear reactions in the primordial fireball need not be followed here, but there are interesting points to be mentioned because they led to exciting work linking nuclear physics with astrophysics and cosmology. The fireball started off at temperatures so high that even the particles making up atoms in the present day universe were not present. The mixture of particles and radiation has been nicknamed the "primordial soup". As it cooled down by expansion two of the principal particles making up atomic nuclei, the protons and the neutrons, formed in virtually equal numbers. Neutrons are stable when they are inside nuclei and bound to protons, but a free neutron on its own is not stable. It decays to a proton, an electron, and a neutrino, in an average time of 880 seconds. While the temperature in the fireball was still high enough, every time a neutron decayed a proton an electron and a neutrino merged to replace it. The number density of the particles was so high and their thermal velocities so great, that the two opposing reactions occurred at equal rates. But as the temperature fell the rate of the backward reaction fell and so the fraction of neutrons grew steadily less. In the same time period the protons and neutrons were beginning to merge to form helium. If the numbers of neutrons and protons had stayed similar most of them would have gone into forming helium, and there would be little hydrogen in the universe. But the number of neutrons fell by decay, and so the proportion of helium nuclei is quite moderate. The calculations suggest that just over 20% of the mass in the primordial fireball became helium and the rest remained as hydrogen. Much smaller fractions were left as deuterium and ^{3}He because those nuclei are not very stable, and most would disintegrate soon after forming. An even tinier fraction went into forming the next lightest element, lithium, in its more stable form as ^{7}Li with three protons and four neutrons. Most of the nuclear reactions deemed to occur in the primordial fireball during the first few minutes after the big bang can be studied in accelerator laboratories, as they are not ultra high energy reactions. They were occurring in the range below tens of millions of degrees K, whereas in its initial instant the temperature of the fireball was very much higher, over 10^{32} K. As a result of these laboratory studies it was possible to predict the values of the primordial element abundances in terms of measured properties of the particles involved.

Astrophysicists picked up this line of research and in the decades between the publication of the discovery of the microwave background and the end of the twentieth century carried out research linking the Big Bang and basic nuclear and particle physics. The first step was to make detailed predictions of how much ^{4}He should have been produced in the primordial fireball, and match these to observations.

The calculations showed that the ratio of ^{4}He to H produced in the fireball should depend on three measurable parameters: the ratio η of the number of baryons (protons plus neutrons) in the universe to the number of photons, the half-life of free neutrons, $\tau_{1/2}$ and the number of different types of neutrinos involved in the reactions. It may seem impossible to measure η, but making an estimate is easier than you might think. Most of the photons in the universe are in fact in the cosmic microwave background radiation. They are low energy photons, because the background is currently at a temperature of only 3 K above absolute zero, but they must all have been present in the fireball at much higher energies, so the number of these photons is the important figure, and they easily outnumber the photons later produced by all the stars in their galaxies. So what the astronomers had to do to measure η was first to calculate the number of photons per unit volume in radiation at the measured temperature of close to 3 K. A very accurate value of this temperature was found by the COBE satellite and published in 1992, but previous estimates were good enough for the basic estimate in photons per cubic metre. The next step was to estimate the mass of baryons in a large enough volume of space to be representative, integrating the masses of galaxies over a volume of diameter several million light years, and finding the average number density, baryons per cubic metre. The ratio can then be calculated. Admittedly this is not a very accurate measurement, because components can be overlooked either due to absorption of starlight by interstellar dust, or by omission of the hottest intergalactic gas which emits only in X-rays, but a "ballpark range" for η of between 10^{-10} and 10^{-9} is found which fixes the range of parameters. We will see below what this ratio means for particle physics.

Knowing the range of values of η the theorists could then predict how much ^{4}He, ^{3}He, D, and ^{7}Li should have been produced in the primordial fireball. A set of these predictions made in the 1980s is shown in Fig. 10.1.

In order to use the predictions we first see that they involve three quantities which are in principle measurable, η, the baryon to photon ratio, the half-life of the neutron $\tau_{1/2}$ and the number of different types of neutrinos N_ν. But these quantities were not very well determined. We can also see that the

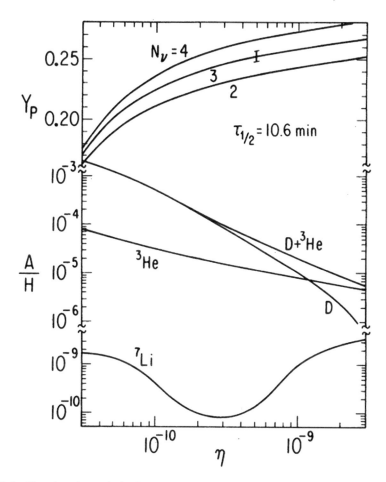

Fig. 10.1 The abundance (Y_p) of Helium-4 (^{4}He) produced in the primordial fireball, in units of the fraction of the total mass of baryons in the universe, (i.e. between 20% and 25%) and the abundances of Helium-3, (^{3}He), Deuterium, (D), and Lithium-7 (^{7}Li) as fractions of the Hydrogen (H) abundance, predicted from models based on the properties of the nuclei and their interactions measured in the laboratory, all abundances calculated as functions of η, the universal ratio of baryons to photons (between 10^{-10} and 10^{-9}). In this figure the half-life of free neutrons before decay is taken as $\tau_{1/2}$ = 10.6 minutes, from then current laboratory measurements. Note that the predicted value for the Helium-4 abundance also depends on the number of different neutrino flavours N_ν. To see how this type of graph is used, see the text. Credits: Yang et al. 1984, Astrophysical Journal, Vol 281, p. 493

predictions are for quantities which astronomers can try to measure, namely the abundances of the Deuterium, Helium, and Lithium. Of these ^{4}He is by far the most abundant, and it is not exceptionally difficult to measure the ratio of ^{4}He/H, often termed Y, using perfectly conventional optical spectroscopy of stars, but more accurately of the interstellar medium. Astronomers knew how to do this, but they also knew that there is a problem. In any part of the contemporary universe the helium we measure is not only primordial, it has an admixture of the helium which is being produced all the time inside stars, and which has been progressively thrown into space either by the winds of "living" stars, or by the explosions of dying stars. There is no magic label which tells us that a particular Helium nucleus was made in the primordial fireball rather than subsequently in stars. Figure 10.1 tells us that unless we measure Y_p, the primordial value of Y, quite accurately it will not give us good information about the value of η, for example. A small change in the measurement of Y_p leads to a big change in the value of η.

The way the astronomers resolved this problem, and essentially the method is the same for ^4He, ^3He, D, and ^7Li is as follows. All the heavier elements, carbon, nitrogen, oxygen, sulphur, and so on were produced in stars. As the universe progressed in time new generations of stars were born, synthesised the elements, including helium and the heavy elements, and distributed them into space. The synthesised elements from one generation were mixed with the interstellar hydrogen making new stars, so that the next generation had a little extra heavy element abundance, more carbon and so on. The most recently formed stars should, in general be the most enriched with heavy elements. Astronomers call the proprtion of elements heavier than helium in any astronomical object its "metallicity" which is a specialised use and does not refer to what we usually think of as metals. In this case they measured two quantities in a set of well chosen stars and interstellar ionised clouds: the general metallicity Z and the helium abundance, Y. They then plotted Y against Z and found the value of Y for which Z has fallen to zero, because the heavier elements were not produced in the primordial fireball, so the earliest stars or clouds should not have measurable metallicity. There are many pitfalls in this process, to the point where there is still significant uncertainty in the results, but essentially the value of Y_p should be the value of Y for objects with the smallest measured metallicity.

Figure 10.2 shows how this works. The observed value is in a range close to 24%, and this is good enough to show that "Big Bang" nucleosynthesis does give a reasonable explanation of the observed ^4He in the universe in general, because as you can see from Fig. 10.1 this puts the value of η somewhere in the middle of the range from 10^{-10} to 10^{-9}, which we could estimate from measurements of the cosmic background radiation and the average local density of baryons in stars and the interstellar gas.

But the possibility of also measuring the primordial abundances of Deuterium, ^3He and ^7Li made the astronomers push forward to see if they

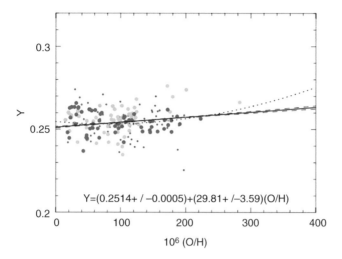

Fig. 10.2 The measured ^{4}He abundance, Y, in astronomical objects with a range of metallicities. The primordial abundance Y_p is found by extrapolating the plot back to zero metallicity. From Izotov et al. 2013 Credits: Izotov et al. 2018 Astronomy & Astrophysics, Vol. 558, p. 571

could use the primordial abundances to say more about the universe. It is much more difficult to measure these three abundances, because they are very small compared to that of Helium-4. The models for Big Bang Deuterium production gave values of order 1 part in 10,000 compared to the Hydrogen abundance, for ^{3}He the ratio is between 1 part in 10,000 and 1 part in 100,000, while for ^{7}Li the ratio is between 1 part in 1000 million and 1 part in 10 thousand million. In spite of these small values astronomical spectroscopy can tackle the problem, using very high resolution spectrographs on large telescopes. One of the conclusions reached, during the 1980s was that the results for all of these light elements taken together did NOT give a common value for η, which was at first seen as disappointing. But it became clear that the discrepancies could be greatly reduced by changing the value for the half-life of the neutron $\tau_{1/2}$. This was initially thought to be unlikely because it was a measurement made in the laboratory and, unlike astronomical data, subject to experimental control. However the astronomers suggested to their nuclear physicist colleagues that it might be worth new experiments to improve the measurement, with the result that the previous value was found to be significantly in error. Measurements before 1980 had found values close to 11 min, and the new values close to 10.6 min gave much better agreement between the values of η needed to get better agreement between the results found from the Big Bang abundances of the different light elements. This was the first case of a purely astrophysical measurement correcting a fundamental measurement in laboratory physics. Experimentalists continue to measure the half-life of the neutron with increasing precision. The current value is close to 10.3 min, but a new and interesting problem has arisen. There are two methods, in one of them a beam of neutrons is fired at a target, and the decay time is measured by the number of reactions per second due to protons from the decaying neutrons reaching the target. The other is to store very cold neutrons in a magnetic trap, the "bottle" method, and count the number of neutrons as a function of time. The half-life measured by the beam method is method 6.5 seconds longer than that measured by the bottle method, and the expected errors in each are around 1 second. The bottle method counts the neutrons as they decay, and the beam method counts the protons they decay into. One explanation for the discrepancy is that some of the neutrons do not decay into protons but into something else, which is not predicted by standard physics. This something else might just possibly be the dark matter particles sought in astrophysics. This example shows the intimate relation between cosmology and particle physics, which forms the subject of this chapter.

From Fig. 10.1 we see that the primordial ^{4}He abundance also depends on N_ν the number of different types of neutrinos in the nuclear reactions. Combinations of particle theory and experiment suggested that this value might be 3, and it was hoped that the observations of the primordial light elements might resolve this question. In the 1980s some of us thought that by measuring the ^{7}Li abundance, which was then the least certain of the four, we might resolve the question, and make a real contribution to particle physics through purely astrophysical measurements. But we were too optimistic. It turns out that the primordial ^{7}Li abundance is very difficult to pin down well by observations, because ^{7}Li was not created only in the primordial fireball, but also in certain types of stars, but worse it is also destroyed in some types of stars, so to isolate the primordial contribution is very complicated.

10.2   The Baryon Density of the Universe and Dark Matter Candidates

While discussing the production of the elements in the primordial fireball we should not forget that the key parameter being measured is η the baryon to photon ratio in the universe as a whole. The light elements do allow us to measure a range for η, around a value of 5×10^{-9}. The energy density of the photons in the cosmic background radiation has been well measured, so we have a value for the number of these photons per unit volume in the universe. From η we can convert to the number of baryons per unit volume, and as each baryon is either a proton or a neutron (with only a small difference of mass between them) we can work out the density of mass per unit volume in baryons. This is a number of great interest to cosmologists. The equations of general relativity of Einstein adapted to the dynamics of the universe as a whole by Robertson and Walker allow us to calculate a value for the density needed for the universe to be "flat", so that light rays propagate along straight lines in a classical Euclidian space. Observations of the expanding universe, and of the cosmic microwave background radiation tell us that the universe is either flat or very close to flat. But the measured value of η shows that the baryon density of the universe is far too small to make it flat. It is less than 1/20th of the required density. This measurement alone is enough for us to question whether baryons, the major well-known components of stars and galaxies, are in fact the really major components of the universe. We learned in the chapter on optical astronomy how the astronomers inferred that the rotation of the galaxies, and the dynamics of galaxy clusters require the presence of a massive unknown component, which we now term "dark matter". At the same time cosmologists and particle physicists were drawing a similar conclusion from very different type of evidence. We now think that even dark matter itself is not the major component of the mass of the universe, and I will touch briefly on dark energy later in this chapter. However the conclusion about the baryonic density, based on η, is quite powerful. It means that no more than some 5% of the mass of the universe could be baryons. This excludes some of the proposals for the constituents of dark matter.

During more than 25 years there were two main types of contenders for dark matter, called WIMPS (Weakly Interacting Massive Particles) and MACHOS (Massive Astrophysical Compact Halo Objects). The former are supposed to be sub-atomic particles whose possible existence had already been hinted at within particle physics. The latter are supposed to be of macroscopic size, such as planetary sized or substellar sized bodies. Searches were made for MACHOS by looking for their effect as gravitational lenses on the images of distant stars, for example stars in the Magellanic clouds if we wish to sample the halo of our Galaxy. A MACHO transiting the line of sight to a distant star should produce a characterisitic rise to a peak and symmetric fall of the light from the star, and an example from one of the MACHO search projects is shown in Fig. 10.3.

These MACHO searches ran for well over a decade, and sampled millions of star images. They were interesting because they did find some very nice light curves for individual events, in the Galaxy and in the Magellanic clouds,

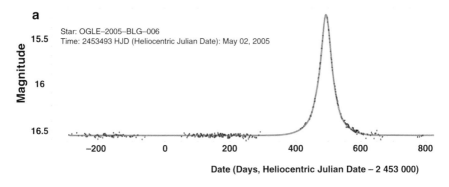

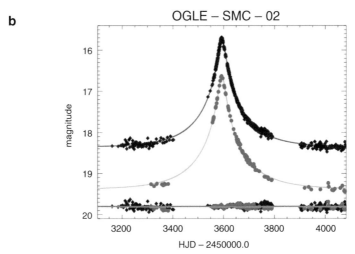

Fig. 10.3 (a) A microlensing event, caused by an asteroid sized solid body passing between the observer and a distant star. (b) A microlensing event caused by the transit of a probable binary black hole observed in two wavelength bands near infrared I band (black points over red model curve) and visible V band (blue points over green model curve). Credits: The OGLE project

including one in which a possible binary stellar mass black hole pair transited across a backgound star. But they did not find a major population, in terms of total mass, of free floating planets, or asteroids, substellar objects or faint stars, or indeed any type of macroscopic objects. This is not suprising because astronomers already knew that baryons make up at most 5% of the mass of the universe, and the dynamical measurements of galaxy rotation showed that a galaxy has five times more dark matter than baryonic matter. So the only known possibility for MACHOS would have been a large population of stellar mass black holes. The MACHO searches ruled out a large population of these, leaving the field clear for WIMPS as the more likely explanation for dark matter. However the MACHO projects have proceded with their long term programme of observing the effects of small bodies transiting stellar sources, which is of especial importance for the properties of extrasolar planets, The gravitationally lensed light curves of these objects are now reguarly used to characterise them, often in conjunction with photometric satellite data from orbiting observatories such as Kepler and TESS.

10.3 Neutrino Flavours

The existence and the properties of the neutrino have been discussed in Chap. 7. It was first postulated by Wofgang Pauli in 1930 to account for conservation of momentum in the decay of certain nuclei. The particle he predicted is now known as an antineutrino, and as a full classification an *electron antineutrino.* We have seen how it was first detected by Cowan and Reines using a nuclear reactor in 1965 for which Reines was awarded the Nobel Prize forty years later. We cannot go at all deeply into elementary particle properties here, but one of the key properties of any particle is its spin. This can be considered by analogy with macroscopic physics, as rotation about an axis. At particle level the spin of an entity is quantized, it can exist only in small finite quantities. The spin of an electron is ½ in these units of quantization, and so is the spin of a neutrino. This applies to both negatively and positively charged electrons (positrons) and to both neutrinos and antineutrinos. In 1936 a "heavy electron" was found in cosmic rays, by Anderson and colleagues. This was named the muon. It is over 100 times as heavy as an electron, and although it is observed to participate in particle reactions of several types, it is unstable, with a mean lifetime of 2 microseconds. Muons have the same charge as electrons, have spin ½ and have their own antiparticles: antimuons. They also have corresponding neutrinos: muon neutrinos, and muon antineutrinos. The third, and up to now final heavy analogue to the electron is the tau, which was found in experiments in the Stanford Linear Accelerator around 1975. The tau is almost 4000 times as heavy as an electron, and decays very rapidly with a mean lifetime of only 3×10^{-13} s. It also has spin ½ and its own neutrino, the tau neutrino. Both the tau and the tau neutrino have antiparticles. The family of particles with spin ½ is termed the lepton family. The electrons, muons and taus all have single units of electrical charge while all the neutrinos are, as their name implies neutral. Each charged particle and its associated neutrino is labelled as a flavour. So there are three charged lepton flavours, and three corresponding neutrino flavours. All particles with spin ½ obey certain statistical rules in quantum mechanics; these rules are called the

Fermi-Dirac statistics and particles which obey them are collectively termed Fermions. Much of this highly compressed summary may not be strictly needed to follow the subject of Big Bang nucleosynthesis but it gives the context, which can be followed up if you are interested.

The equations which yield the fraction of different elements and their isotopes produced in the primordial fireball contain reactions whose overall speed depends on the availability of neutrinos to convey momentum and spin, and specifically on the number of different neutrino flavours. When Fowler Wagoner and Hoyle published their first paper on Big Bang nucleosynthesis it was not clear how many flavours exist, and theorists made calculations for values of 1, 2, and 3, as shown in Fig. 10.1. Even now fundamental particle physics does not set a final limit of three to the number of neutrino flavours. When observational astronomers set out on their search to measure the primordial abundances of ^{4}He, D, ^{3}He and ^{7}Li they realised that one of their aims was to see if they could produce a "concordance" set of values for η, the baryon to photon ratio in the universe, $\tau_{1/2}$ the of the average lifetime of the neutron, and N_ν the number of neutrino flavours: four measurable parameters for only three unknown quantities. They appeared to come close to achieving this as early as the 1980s. We have seen how they were able to oblige the laboratory physicists to re-measure the neutron lifetime, and another of their conclusions was an upper limit to the number of neutrino flavours of four, with measurements pushing this tantalisingly close to three, which would have meant a fundamental contribution to particle physics. However the weak observational link was, as we have mentioned, the primordial Li abundance, and as time has gone on the uncertainty about measuring the true primordial abundance of ^{7}Li has increased. The importance of neutrinos in astronomy in general has grown in recent years to the point where neutrinos are now clearly independent messengers in the panoply of multi-messenger astronomy, which we have already dealt with in another chapter. In the meantime measurements of the parameters of the microwave background radiation have become increasingly accurate, and are being used to throw more light on the links between particle physics and the Big Bang.

10.4   The Cosmic Microwave Background and Precision Cosmology

In the chapter on radioastronomy we saw how the background of microwave radiation which emanated from the primordial fireball, and fills the universe today was first measured at centimetre wavelengths by the radioastronomers and how subsequent ground based and space based observations have, since its discovery over 50 years ago, gone on to increase our knowledge of its properties. Here we will look at some of those properties to see how they allow us to understand the universe on its largest scale and also link with unsolved problems in particle physics.

In Fig. 10.4 we show one of the two observations made in 1992 by the Cosmic Background Explorer (COBE), the averaged spectrum of the background over the whole sky, taken with the FIRAS spectrometer. This observation is quite remarkable, in that it shows that the spectrum of the CMB is indistinguishable from the theoretical spectrum of a black body, an ideal physical concept predicted as being the most probable spectrum for the fireball radiation. The precision of this observation, and of the inferred temperature of the radiation (2.725 K above absolute zero) is a tribute to the team on

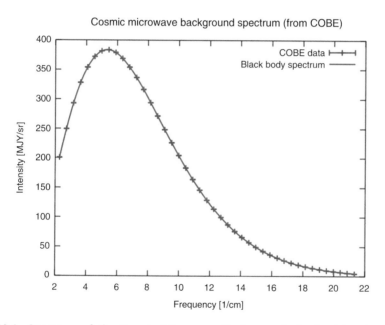

Fig. 10.4 Spectrum of the Cosmic Microwave Background obtained by the FIRAS experiment on the Cosmic Background Explorer (COBE) satellite, compared with the theoretical spectrum of a perfect black body. The error bars on the measurement are smaller than the width of the observational plot and cannot be represented on the figure. The agreement between observation and theory is astonishing. The spectrum gives a temperature of the background radiation of 2.725 K above absolute zero, which corresponds to −270.419 C. The units of the Frequency axis are inverse centimetres. This means that 10 units corresponds to 1 mm wavelength, and that the peak of the spectrum is close to 5.5 units, which is 1.8 mm. Credits: NASA/The COBE project research team

the experiment led by John Mather, who received the Nobel Prize in 2006 for this work. Nobody has thought seriously of trying to improve the measurement, which is the bedrock of modern observational cosmology. The other main experiment on COBE was the DMR, or differential microwave radiometers, at frequencies of 31, 53 and 90 GHz, corresponding to 9.6 mm, 5.7 mm and 2.33 mm wavelengths respectively. This was designed to measure the variation in intensity of the CMB from place to place on the sky, which from ground based measurements was already known to be small.

In Fig. 10.5 we show a simplified summary of the results of the DMR. The upper picture is just the average temperature of the whole sky, so by definition it has a uniform value, set by calibrating it by the FIRAS result, at 2.725 K. The second picture is obtained by subtracting off this value from the whole sky map, but with very great care to mask off the emission from the galactic plane zone, and smoothing the result for a resolution of 10 degrees on the sky. This map shows a simple pattern, called the dipole pattern, which had in fact been found previously in measurements from aircraft and stratospheric balloons. This dipole distribution of brightness is quite a small ripple on the underlying background radiation. It has a temperature amplitude ΔT of 3.3 mK, in the millikelvin range, not much more than one thousandth of the value of the basic temperature. It is produced because we have a velocity in a given direction in space with respect to the background radiation, and the DMR result showed this direction to be, in equatorial co-ordinates: Right Ascension 168.9° (± 0.5°) Declination −7.5° (± 0.5°) (these coordinates are an equivalent on the sky to longitude and latitude). This motion is thought to be due to the gravitational effect of a large scale mass concentration, nicknamed the Great Attractor, postulated to be a supercluster of galaxies in the direction of the constellation of Vela. The third and lowest map shows the result of removing the uniform background, and the dipole component. What is left is a residual of the emission from the Milky Way, concentrated towards the plane, plus a set of zones with slightly higher or slightly lower temperatures. We can see the scale of ΔT, the amplitude of the variations across the sky, as 18 µK, in microkelvins, hundreds of times lower, even, than the dipole component. The variations in map three were an early hint of the presence of the fluctuations, or anisotropies, in the CMB, expected to be present in any Big Bang scenario showing the imprint of density variations which eventually gave rise to galaxies and stars. They had been sought for over a decade. The DRM made the first detections of these anisotropies at three different wavelengths, and the Principal Investigator on the DMR, George Smoot, was awarded the Nobel Prize for this. But the amplitudes of the anisotropies are so small that a significant fraction of what was first published from the DMR result was noise

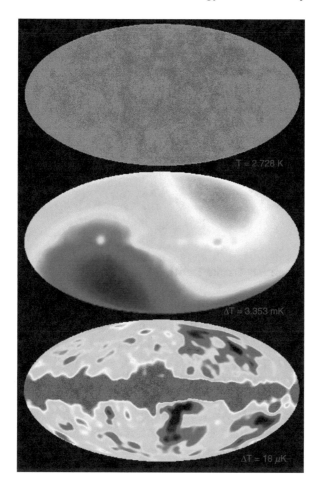

Fig. 10.5 Maps of the millimetre wave sky from the COBE differential microwave radiometers (DRM) published in 1992. The upper map just shows the mean temperature of the CMB radiation, in good agreement with the measurement by the FIRAS spectrograph, also on COBE. The middle map shows the dipole component of the CMB, due to our motion with respect to the radiation field. The lower map shows the anisotropies over the sky, at very low level. Those near the mid-plane of the map are residuals due to emission from the Galaxy, but further from the plane the cosmological fluctuations are present, and detected for the first time with COBE Credits: NASA/The COBE project research team

rather than signal. A deeper map of the anisotropies over an 8° wide strip of sky, was made in the Tenerife experiment from the ground, and published in 1994. Once these anisotropies had been detected a series of experiments from the ground, from stratospheric balloons, and from satellites provided increasing sensitivity and more information about them, and hence about the conditions in the primordial fireball. The major steps were taken with NASA's

Fig. 10.6 Map of the CMB anisotropies by the ESA Planck satellite, complete data from 9 years of operation. Credits: ESA/The Planck project team

WMAP satellite, launched in 2001, and with ESA's Planck satellite, launched in 2009. Planck has provided us with the most recent and deepest maps of the CMB. The Planck mission had a limited lifetime because one of its two instruments, the high frequency instrument HI-Fi, was cooled with liquid helium and the supply ran out in 2012. However the complex data have been analysed and re-analysed and three data releases, in 2013, 2015, and 2108 led to a steady refinement of the results. The map of the anisotropies on the sky and the graph of their temperature distribution as a function of angular size shown in Figs. 10.6 and 10.7 respectively were presented in technical publications in 2018.

10.5   How the Theorists Explain the CMB Anisotropies

Theorists have worked hard, and with their usual imagination, to quantify the interaction of particles and radiation in the primordial fireball and the behaviour of the universe during the almost 400,000 years until the particles and the radiation decoupled from another. In this book, which aims at covering a wide range of ways in which we learn about the universe it cannot be an aim to go into the theory, as superficial treatments of theory tend to mislead. But some description of what is being understood is well worth while. In Fig. 10.7 you can see the observational results from Planck being compared with a theoretical fit to the standard model of cosmology. This was developed from

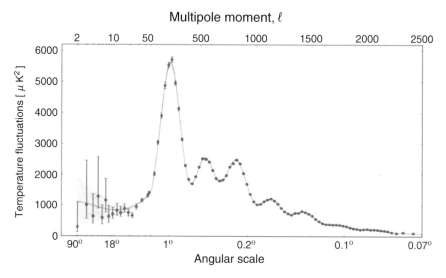

Fig. 10.7 Distribution of the temperature in the map of Fig. 10.6. The vertical axis is the square of the measured temperature of the fluctuations in μK and the bottom horizontal axis is the angular scale, which is linear up to 20° and then logarithmic, to better include all the data. The points are the measured observations from the Planck satellite, using the full final reduced and processed data set up to 2018. The green line is the best fit cosmological model from which up to six essential parameters of the observable universe can be obtained. The agreement between model and observations for small angular scales is outstanding, but at scales bigger than 10° the observations fit much less well. Observational error bars are shown as vertical bars, and model uncertainty is shown as the green shaded area. The upper scale is in units of multipolarity; 2 means a size of a quarter of a circle on the sky, 50 means a size of 1/100 of a full circle on the sky, which is 3.6 °, etc. Credits: ESA/The Planck project team

observations of the expanding universe made using the light from stars and galaxies, and based on the dynamical equations of the general theory of relativity. It then incorporated systematically the properties of the particles and the radiation participating in the expanding universe which we have been able to measure in the laboratory, and the laws of behaviour of large numbers of these particles and photons as they interact with themselves and one another. The small variations in the density of the primordial soup which gave rise to the anisotropies in the CMB, and later to the development of structure in the universe are postulated to be the result of quantum fluctuations in the earliest stages of the universe, in a period known as the inflationary period, and this allowed theorists to predict a spectrum of sizes for the initial fluctuations, and also how these would develop as the universe expanded. This standard model was then used to predict the angular spectrum of temperatures on the sky we observe today, which is the green line in Fig. 10.7. In order to get the numbers, the shapes, and the heights of the peaks and troughs in the diagram the model fits a number of parameters, of which the six most influential are: the density of baryons Ω_B, the density of dark matter Ω_{DM}, and the density of dark energy Ω_{DE}, the current expansion parameter (the Hubble-Lemaître parameter) of the universe H_0, the amplitude of the primordial fluctuations, A_s, and the slope of the spectrum of these fluctuations, given by a parameter n_S. The theory of inflation puts n_S at one or very close to one.

As the aim is to make this chapter as self-contained as possible I will give a brief account of the parameters not already discussed earlier in the chapter, before reporting on the values obtained from the Planck results. Firstly there is the Hubble-Lemaître parameter H_0, the measure of the expansion rate of the universe measured locally in space and time. It has been measured historically using the known expansion velocities and distances of galaxy clusters up to a thousand million light years from us, where the effects of differences in the cosmological models may be considered small. The velocities can be measured quite easily using the Doppler effect in their spectra, but the distances have always been a problem. There is a huge quantity of work on distance measurements in astronomy, as any valid physics applied to stars and galaxies needs them. The principle of measurement entails finding "standard candles", objects whose absolute luminous power is known from their basic physics, so that by measuring how much light we receive from them we know their distances. The two most useful standard candles are the Cepheid variable stars, for local distance measurements, and type Ia supernovae for much greater distances, reaching out to cosmologically important distances. After almost a century of study, with increasingly refined and verified techniques, involving the Hubble Space Telescope for the Cepheids and many large ground based telescopes for the type Ia supernovae astronomers agreed on a value for H_0 of 73 ($\pm$ 1.4) km/s/Mpc, which means that for every additional megaparsec of distance separation between two clusters of galaxies the velocity at which they are separating is an additional 73 km/s. Three further parameters which can be used to fit the power spectrum of the temperature fluctuations of the CMB are three densities, which we have touched on previously but which I will state clearly here. One is the universal mean baryon density Ω_b, which primordial nucleosynthesis has shown to be some 5% of the density needed just to close the universe, the second is the density of dark matter Ω_{DM} which studies of the dynamics of individual galaxies and clusters of galaxies put at around 5 times that of the baryon density, that is 25% of the density needed to close the universe. The third density Ω_{DE} is that of the dark energy, which is deemed responsible for causing the universe to accelerate. This acceleration was inferred from measurements of velocities and distances of galaxy clusters on the largest measurable scales, using type Ia supernovae as the distance indicators. Its revolutionary consequences for our understanding of the universe were recognised by awarding the Nobel Prize to three of the astronomers on the two teams which made the discovery, Saul Perlmutter, Adam Riess, and Brian Schmidt. The mass equivalent to this dark energy can be assumed to provide the mass which closes the universe, provided that it is indeed just closed, but the observations of the expansion do not give direct evidence for this. Another of the six parameters which can be fitted by the cosmological models to the observations of the CMB temperature anisotropies in Fig. 10.7 is n_s the the index of the primordial fluctuations in density, which were the seeds of all the present structure. This index would have the value 1 if the power in the fluctuations were independent of the scale, with fluctuations on all scales being equally probable. If the origin of the fluctuations was just quantum uncertainty at the instant of the Big Bang this index should be 1 or very close to 1. The sixth parameter which can be varied to achieve the best fit by theory to the observations by PLANCK is the amplitude A_S of the initial fluctuations. The power spectrum shown in Fig. 10.7 has a sufficient number of well defined peaks that more than six observable parameters could in principle be fitted and hence derived from this observation alone. These include the mean temperature of the CBM radiation, and the number of different neutrino flavours N_n which affected the conditions within the primordial fireball. The CMB temperature was found with impressive accuracy by the spectrum experiment in COBE and can be used as a fixed parameter, an input to the interpetation of the fluctuations. The number of neutrino flavours as predicted by the standard model of particle physics is 3, and this too is normally taken as an input to the interpretation of Fig. 10.7.

To interpet the observations in the power spectrum the first step is to measure the position of the first, highest peak in the figure. We can see that this is at just under 1^0 and theory gives this a very direct intepetation, that the universe has a flat geometry, which implies that the total density has the value required just to close the universe. From that point a full set of parameters is used to produce the fit to the data shown in the figure, and the best fit gives a set of values for the real universe. The resulting values of most interest to us are: H_0, the current parameter for the expansion rate of the universe, is 67.7 ($\pm$0.4) km/s/Mpc, Ω_b, the baryon density of the universe, is 5.2 ($\pm$0.05) % of the closure density, Ω_{DM}; the dark matter density of the universe is 26 ($\pm$0.2) % of the closure density, and the age of the universe is 13.79 ($\pm$0.02) Gyr. These are values with very small apparent uncertainty, which has given the description of precision cosmology to the subject. As the universe is shown to be flat, this means that the total density Ω_0 must take the value one, i.e. 100%.

So from the cosmic background radiation alone, without requiring any other values, we can see that if only 31.2% of the matter in the universe is accounted for by the sum of baryonic and dark matter, the other 68.8% must be in some other form.

Before this precision work on the CMB had been performed, the acceleration of the universe had been demonstrated by the observations of supernovae type Ia in distant galaxies, as we have already seen, and the existence of a component of energy to cause this acceleration had been postulated. It was formulated as an extra term in Einstein's equations of General Relativity, the Λ term, which he had introduced as a way of avoiding the contraction of the universe under its own gravity, before its expansion had been discovered by Hubble. After the expansion was demonstrated, Einstein made the famous statement that the introduction of the Λ term had been "my greatest blunder" but when the accelerating universe was revealed in 1998/1999 this term was opportunely in place to describe it. To the point that the now standard scenario for cosmology is called the Λ Cold Dark Matter, usually shortened to ΛCDM, model. The fact that the baryon density predicted by the CMB radiation alone has the same value as that previously predicted by Big Bang nucleosynthesis, with completely different physics involved, is a powerful argument in favour of the ΛCDM scenario. The fact that the dark matter density obtained from the CMB radiation agrees with that predicted from the dynamics of individual galaxies and of galaxies in groups offers a further argument in its favour. However until the particle physicists find a plausible candidate for dark matter particles, alternative explanations are open.

One point which you may have noticed is that the Hubble-Lemaître parameter, H_0, has two different measured values, one obtained using the expansion rates of the galaxies, which I have quoted as 73 (±1.4) km/s/Mpc, and the other obtained from the CMB radiation: 67.7 (±0.4) km/s/Mpc. The first of these measurements has been refined by new observations since I initially wrote this chapter, and best estimates put it close to 71 km/s/Mpc. Nevertheless the two values are considered to be too far apart to be explained simply on the basis of observational uncertainties. The experts in the field talk about an "unresolved tensión" between them, which could indicate new physics or cosmology yet to be revealed.

There is quite a lot more physics in the Cosmic Microwave Background, of which I can offer only a fleeting description. The theory of inflation predicts that in addition to the sound waves in the primordial plasma which caused the peaks in the power spectrum there should also have been gravitational waves emitted. These could in principle be detected via specific patterns in the polarisation of the CMB, but they are fairly weak patterns. Stronger patterns of polarisation associated with the fluctuations we have been talking about have been well measured by the PLANCK satellite, and they are used to improve the already good values of the cosmic parameters described above. But the different, weaker patterns, the so-called B modes, have not yet been clearly observed. Their observation would clinch the already strong arguments in favour of the inflationary model of the earliest epoch in the observable universe, and they are being eagerly sought via specialised ground based telescopes.

10.6 The Big Bang and Particle Antiparticle Asymmetry

Although all the types of sub-atomic particles have their antiparticle counterparts we live in a world where particles dominate. We know that particles and their antiparticles annihilate when they meet, producing gamma-rays. In the primordial fireball the huge temperatures and densities ensured equilibrium such that particle antiparticle pairs were being produced and annihilated at virtually the same rate, but as the fireball cooled and expanded the probability of annihilation predominated. However an asymmetry in their properties led to a small numerical excess of particles over antiparticles, so that the final result was a small remnant of particles in a sea of gamma rays. We know the value of the small excess because it is coded in η, the ratio of particles to photons in the universe, which we have seen takes a value close to 1 part in $10^{9.5}$. This is just the value of the tiny proportional excess of particles over antiparticles, and we need hardly emphasise its importance. We are formed from this excess. But it leaves the question of what properties we can link to the asymmetry which would give a clue to its cause. While writing this chapter a possible step towards understanding this was taken at the Super-Kamiokande neutrino observatory in Japan, which figures in our account of neutrino astronomy in Chap. 7. In an experiment labelled T2K an international team of scientists created a beam of muon flavoured neutrinos at Tokai, and fired them at the Super-Kamiokande detector 295 km away. We saw in our neutrino chapter how neutrinos oscillate spontaneously between their three different flavours. The physicists were looking for neutrinos and antineutrinos which had oscillated from muon flavour to electron flavour during their flight from Tokai to Super-Kamiodande. The standard particle physics model predicted that the ratio of neutrinos to antineutrinos finally detected should be 68 to 20 but the result was a ratio of 90 to 15. This implies that neutrino oscillations have a higher probability than antineutrino oscillations, which is a particular case of a general property called CP-violation. How does this finding relate to the asymmetry between matter and antimatter in the Big Bang? The particle physicists postulate that neutrinos must have a counterpart with much higher mass, rather the way electrons have muon and tau counterparts with much higher masses. These higher mass neutrino and antineutrino counterparts could have formed only in the extreme conditions of temperature and density within the primordial fireball. If they also showed CP violation during their decay this could give rise to the matter antimatter asymmetry, leading to the present predominance of matter, and the virtual absence of antimatter. It is fair to say, though, that the Japanese experiment we are discussing here, and other experiments with similar aims in other countries, have not yet accumulated sufficient data to be able to claim a real proof of CP-violation for neutrinos. This is probably not too far off in time, but any confirmation of the existence of the heavy neutrino and antineutrino counterparts remains experimentally complex, and unlikely to occur soon.

10.7   Dark Matter Searches

Dark matter is a fully satisfactory component of the models which explain the CMB anisotropies, and at the same time it is required to explain the internal dynamics of galaxies and clusters. Particle theorists had proposed the existence of particles with significant mass and only gravitational interaction with the rest of matter before these astronomical and cosmological considerations were thought to be needed. Nevertheless science cannot maintain indefinitely a hypothesis without direct evidence. This has given the impulse to a variety of different ways to search for dark matter particles. In the chapter on gamma-rays we saw how astronomers have looked for the gamma rays produced by a dark matter decay or annihilation in the suggested dark matter density concentration at the centre of the Milky Way, with so far no detection. Laboratory tests for dark matter have been multiplying since the 1980s. These tests are sited in underground laboratories in order to minimise the possible confusion of dark matter particles with cosmic rays or cosmic ray induced particles. There are over 20 underground experiments at European, Asian and American sites, and also in Antartica. The basic technique is to look for the recoils at quite low energies (in the keV range) caused by the interaction of a dark matter particle with a nucleus. There are two types of detectors in use, ultra-low temperature (cryogenic) detectors and liquefied noble gas detectors. The cryogenic detectors are looking for heat emitted when a dark matter particle hits a nucleus in a crystal such as germanium held at a temperature below 100 millikelvins above absolute zero. Noble liquid detectors detect a light scintillation produced by a particle collision in liquid Xenon, or Argon. Both techniques are designed to distinguish between gamma-rays, which scatter off electrons, and the dark matter particles, which scatter off nuclei. Here I will give a description of only one of the Xenon experiments, XENON1T currently the most sensitive. It is depicted in Fig. 10.8.

This is sited in one of the world's best known underground physics laboratories within the mountain at Gran Sasso, Italy, 1400 metres below the ground. The detector uses three tons of ultra-pure liquid xenón, at –110 ° C. in a cylindrical vessel, with Xenon gas above the liquid. It has an array of photomultiplier detectors below the liquid and another array above the gas. Gamma rays interacting with the electrons in the xenon atoms produce UV photons at 175 nm, which travel to the phototube arrays above and below the tank, and allow the interaction point to be located in 3 dimensions, using the times of flight to the two arrays, and the positions of the phototubes which make the detections. This signal is termed S1. Electrons generated by the interaction are moved upwards into the gas by an applied electric field, and generate photons by electroluminescence, which are also detected by both upper and lower phototube arrays. This signal is termed S2 and arrives later than S1. The WIMPS, the dark matter particles being sought, would interact only with the XENON nuclei. They would also produce signals S1 and S2. The difference between the two types of interactions is that the predicted strength ratio S1:S2 for WIMPS is much bigger than the predicted ratio for gamma-rays. Neutrons interacting with the liquid xenon will produce a similar ratio of the signals to that expected from WIMPS. However neutron interactions are strong, and one would expect several interactions from the same neutron, whereas WIMP interactions are weak and only one per WIMP is predicted. Experiments such as XENON1T look for single interactions. This simplified description of a laboratory method for WIMP searching has not included the details of the complex shielding needed for the XENON liquid vessel to avoid entry by other types of particles, above all muons produced by cosmic gamma-ray reactions. The XENON collaboration involves physicists from 21 institutions in 10 countries. It is in continual renewal and improvement; its detector chamber will shortly be filled with 10 tonnes of the ultra pure liquid xenon. It is sensitive to WIMPS with hypothesised spin, as well as without spin. However so far it has not detected any WiMPs, nor have any of the other relevant experiments in underground laboratories.

Another way to look for dark matter particles is to use a particle accelerator to produce them in situ. The most powerful accelerator at present is the Large

a

b

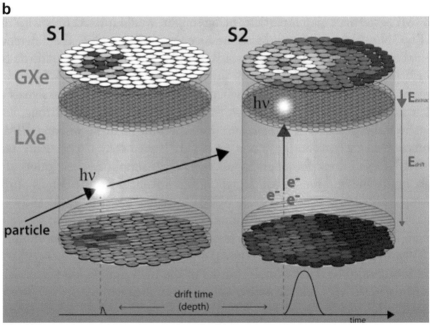

Fig. 10.8 (a, upper). Left: The liquid Xenon chamber of the XENONnT dark matter search detector at the underground Gran Sasso Laboratory in Italy. Right: The auxiliary 3 storey building with the machinary for purifying the Xenon in the upper Storey, a distillation plant to separate Xenon from Krypton in the lower Storey, and the elec-

Hadron Collider (LHC) at the CERN particle physics laboratory, Geneva, Switzerland. There are two experimental set-ups at the LHC, ATLAS and CMS, which examine the products of the high energy collisions between protons produced in the collider, and a dark matter study group has been set up to search for evidence of dark matter particles in the records of these products. From the point of view of processes in the universe as a whole, for instance those which produce cosmic rays and gamma-rays, the "high energies" of the particles produced at the LHC are quite low, but as there is no clear prediction of the masses of the dark matter candidates, they could be anywhere between 0.1 GeV and 10 TeV, the large flow of data produced enables this low energy range to be surveyed with considerable discrimination. The method used is not basically different from that used in previous successful searches for particles within the standard model of particle physics, such as the W, the Z, and the Higgs bosons. The collision products are examined, all their energies, momenta, and directions are measured, and anomalies not explicable in terms of known particles and their known decay products are examined and explained.

The results, up until the date of writing this, in mid-2020, have not produced any evidence for any particle beyond those in the standard model. It is important to note that there are extensions to the standard model which do predict the existence of particles which could comprise dark matter, so these experiments are of considerable interest to particle physicists as well as to astronomers.

The third place to look for dark matter is in the particle and gamma-ray showers produced by the high energy cosmic rays after they hit the upper atmosphere. We have described in Chap. 9 how these measurements are made in general with the large ground-based telescopes such as the Cherenkov Telescope Array (CTA). We have also mentioned an indirect way to look for dark matter using these telescopes, which is to try to observe the gamma-ray signals from the annihilation of dark matter particles in those places where we expect maximum densities of these particles. This means in particular the centre of the Milky Way, and also the centres of nearby dwarf galaxies, as their internal kinematics point to higher ratios of dark matter to baryonic matter,

---

Fig. 10.8 (continued) tronics for controlling the experiment in the storey between them. (**b**, lower). Schematic diagram of the XENON underground experiment searching for dark matter particles. The two types of signals expected from a WiMP interaction with a Xenon nucleus, S1 and S2, are described in the text. Eneegy deposition in the liquid Xenon causes scintillation signal S1 (left); electrons from the ionization are extracted into the gas phase and amplied, giving signal S2 (right). Circles represent photomultiplier detectors. *Credits: XENON1T collaboration*

and other sources of gamma-rays are usually absent. Other suggested places to look for dark matter signatures are galaxy clusters, as their gravitational fields are also dominated by dark matter. The CTA is still in its infancy, and so far the telescopes working on similar principles, such as MAGIC and HESS have failed to find the relevant gamma-ray signatures. However the CTA will extend the range of particle energies from the GeV to the TeV range, opening possibilities of detecting higher mass particles. An additional indirect way to detect dark matter using gamma-rays is a prediction that the high energy sources which emit gamma-rays could also emit the prospective dark matter particles called axions. The axions can decay into gamma radiation in the presence of magnetic fields. This would boost the high energy part of the gamma ray spectrum of sources emitting strong gamma-ray flares. The spectral distribution of this excess is predictable, and could be used to detect the presence of the original axions. You can see that we are reaching really indirect ways of trying to find dark matter, which is a symptom of the wish to find it, and the anxiety caused by the difficulty. Still we should not forget that neutrinos, whose mass is too small for them to give a significant contribution to the dark matter, were first discovered 26 years after their prediction, so the hunt for dark matter particles should not yet be branded a failure. It will take time, ingenuity, and effort to find them.

Further Reading

The best book ever written on cosmology for the general reader is "The First Three Minutes" by Nobel Laureate in Physics Stephen Weinberg, first published in 1973. Of course there have been major developments since then, but the fundamentals are very well explained. It is available on several websites. A downloadable pdf version can be found at http://slobodni-univerzitet-srbije.org/files/weinberg-steven-the-first-three-minutes.pdf

Peebles, P. J. E. "Cosmology's Century" Princeton University Press. 2020. An excellent book from recent Nobel Laureate Peebles, deeply involved in the subject. But expensive, ask your library to get it! https://press.princeton.edu/books/hard-cover/9780691196022/cosmologys-century

For physics students Ulrich Ellwanger "From the Universe to the Elementary Particles" Springer, 2012 https://www.springer.com/gp/book/9783642243745

For entertainment plus instruction "The Cosmic Revolutionary's Handbook" by Luke Barnes and Geraint Lewis, Cambridge University Press, 2020.

Books by Stephen Hawking. I put these in descending order of recommendation: "The Universe in a Nutshell", "The Grand Design" (with Leonard Mlodinow), "A Briefer History of Time" (with Leonard Mlodinow), "Black Holes and Baby Universes" "A Brief History of Time", all published by Penguin Random House

11

Hands-on Astronomy: Meteorites and Cosmochemistry

11.1 The Use of Material from Outside the Earth for Astronomy

The detection of neutrinos from the Sun and from Supernova 1987a followed by the detection of gravitational waves from merging back holes and neutron stars showed that astronomical observations were possible using other clues than those carried by electromagnetic radiation. The term "multimessenger astronomy" was introduced to characterise this widening of horizons. But there are two other, already traditional, ways of exploring the universe which do not use electromagnetic radiation. We have already seen one of these in the chapter on Cosmic Rays, where particles from high energy astronomical sources are being detected, albeit often through the mediation of gamma-rays. The other is the use of macroscopic material from outer space (often in microscopic quantities!) in the form of meteorites and lunar and cometary samples. This is really "hands on" astrophysics, carried out in laboratories. Much of this information bears on the formation of the solar system, and in particular the formation of the planets within the accretion disc of gas and dust in the equatorial plane of the rotating Solar System. But there is also a growing body of accurate measurement related to the formation of the chemical elements in stars. A specific property of this material, because it allows direct laboratory assays of constituents, is the use of radioactive isotopes to identify timescales and even narrow down the times of the events under study. When related to the results of dating clusters of stars using stellar evolution theory this is a

© Springer Nature Switzerland AG 2021
J. E. Beckman, *Multimessenger Astronomy*, Astronomers' Universe,
https://doi.org/10.1007/978-3-030-68372-6_11

powerful tool to enhance our understanding of evolutionary processes in astrophysics in general.

11.2 Meteorites and What They Can Tell Us

11.2.1 Types of Meteorites

Meteorites falling to Earth at a rate of over a hundred tons per day are by far the most common source of extraterrestrial material. Although stones falling out of the sky have been known for millenia, their nature did not begin to be understood until the turn of the nineteenth century. And it was not until the 1950s and 1960s that a combination of radio and optical observations gave us the answer to their origins. The radio observations were in fact by radar, bouncing a beam of centimetre wavelength radio waves off the ionised tracks left by meteors, a technique pioneered in Sheffield by Tom Kaiser, who showed me his radar laboratory on the moors near the city when I was at gramar school there. The optical methods were photographic. Between the two techniques they could measure the velocity, the speed and the direction, of meteors as they arrive into the atmosphere from space. This was sufficient to show that by far the majority of meteors have their origin in the asteroid belt, between Mars and Jupiter, and have been deviated inwards by interactions with bodies in similar orbits. Although observed meteorite falls are very important, because they can be collected before weathering in our atmosphere, most meteorites are found by chance, and the best sites are in the driest parts of the Earth, notably Antartica and north Africa.

Most meteorites (over 90%) are stony, and only 1% comprise equal parts of metal (iron-nickel mix). The basic classifiaction of stony meteorites is into chondrites, and achondrites, Chondrites have similar composition to the Sun (apart from missing the most volatile elements) and come from asteroids which have not "differentiated", i.e. which have not been melted, separating an iron rich core from a rocky mantle and crust. Achondrites are mainly igneous, having originated on differentiated bodies, which include known origins on the Moon, Mars, and the asteroid Vesta. The chondrites are more primitive; they contain a record of the earliest materials from the solar nebula, not reprocessed on their parent asteroids. The chondrites have three main components (a) chondrules, small spheres, under 1 mm in diameter, formed as droplets which cooled and crystallised in space; (b) refactory inclusions (RI's) with sizes of up to centimetres, of which one type the CAI's, (calcium-aluminium-rich inclusions), have compositions similar to that predicted for the first solids to form from the gas of the solar nebula. They can be dated by radiochronology and are used to define the age of the Solar System (c) the matrix, a fine grained (less than one micrón sized) aggregate of mineral and amorphous particles packed between the chondrules and RI's. It includes broken chondrules and RI's but also trace quantities of presolar dust grains. These are tiny grains which originated in previous generations of stars, and can be recognised by their anomalous composition, especially the anomalous ratios of their isotopes. The matrix also includes organic carbonaceous material which may have come from the Sun's parental molecular cloud. The most primitive meteorites, especially the carbonaceous chondrites, show little evidence for processing, either by heat or by water, and are especially valuable for astrophysical interpretation.

As well as conventional meteorite samples, there are tiny meteorite grains found in Antarctica (AMM's), and dust particles gathered on collectors attached to the wings of aircraft flying in the stratosphere at heights above 17 km which is the upper limit reached by terrestrial dust from below. Both sets of particles can be divided into hydrated (with water molecules included) and anhydrous. It has been suggested that the hydrated particles come from asteroids, and the anhydrous particles come from comets. They are similar to the matrix of the meteorites showing mínimum processing of their parent bodies, with high abundances of carbon and nitrogen, and no evidence of alteration by heat or water. Those from the aircraft samples are termed IDP's, interplanetary dust particles, and they come from the population of particles whose scattered light in the plane of the ecliptic has been called, historically, the zodiacal light.

Chondrites, in their various forms give us most information about the early Solar System, how it formed, and indeed about the material in the Galaxy before its formation. The carbonaceous chondrites, which have quite a small proportion of carbon (but clearly have more than the ordinary chondrites) are those which have experienced the least alteration while being accreted into larger bodies, and so are the most useful, together with the AMM's and the IDP's mentioned just above. Together with the lunar samples from the Apollo project, dust returned from comet Wild 2, and dust returned from asteroid Itokawa these are the most intensively studied pieces of extraterrestrial material.

11.2.2   Pre-solar Grains

One of the key steps to our understanding of how to use meteoritic material in astrophysics was the discovery that it was possible to separate material which had not been affected by the mixing process of the elements which went into the formation of the pre-solar nebula and the Solar System itself some 4.6 Gyr ago. This discovery was based on the presence of anomalies in the ratios of isotopes of the noble gases (mainly xenon and krypton) and also of oxygen in the CAI particles, in the 1960s and 1970s. These anomalies were understood to be due to the preservation, within the pre-solar material, of the imprint of different nuclear processes for synthesising the elements in individual stars. An isotopic anomaly is a ratio of isotopes which cannot be derived from the average found in solar system material via chemical or physical processes, such as evaporation or diffusion. The principle of using isotopes in this way is that the ratio of the isotopes of a given element depends on the nuclear processes which formed the element in a star, and remains almost unchanged during physical aned chemical processes once it has left the star. The "almost" has an important exception. Because deuterium has twice the mass of normal hydrogen, the formation of compounds with deuterium is in fact greatly affected by the physical conditions in which the chemical reactions take place, notably the temperature. The lower the temperature the higher the ratio of deuterated to normal hydrogen molecules of all types. But the heavier the element the smaller is this effect, and in particular even for elements which are not especially heavy, such as carbon, nitrogen, and oxygen, these differentiation effects are small, so that isotopic anomalies can be directly traced to processes within the stars where they were formed.

Figure 11.1 gives a pictured scheme of the kinds of processes linking stars, interstellar gas and dust, and planet formation.

The key points, as far as analysis of meteoritic and cometary material is concerned, are the two principal "dust factories": lower mass stars emit large quantities of dust into their circumstellar media when they are in the red giant phase of evolution, while high mass stars produce and emit dust when they explode as supernovae. There is a specific part of the evolutionary sequence of lower mass stars which have ended their main hydrogen burning phase and are burning helium, having been puffed out to large diameters in the process, called the asymptotic giant branch (AGB) during which they emit large quantities of dust. Because they have big tenuous outer atmospheres, the gravitational field there is low, and they lose a lot of mass in stellar winds. And as they are relatively cool, the elements in the winds condense to form dust grains.

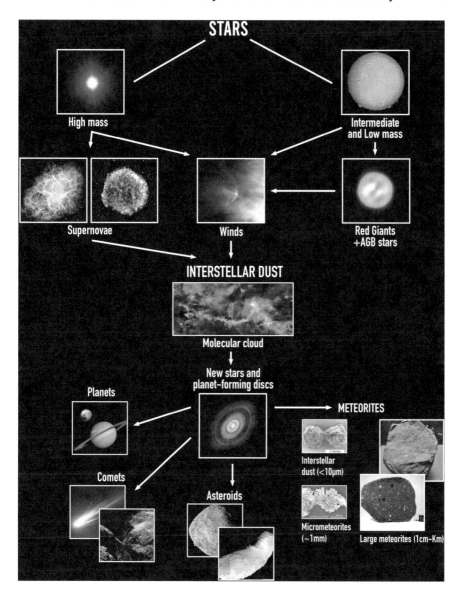

Fig. 11.1 Connections between stellar, interstellar and early planetry processes and extraterrestrial materials studied in terrestrial laboratories. Meteorites, micrometeorites, and Interstellar dust particles are naturally delivered samples of asteroids and comets which are remnants of planet formation in the Sun's protoplanetary disc. These samples contain records not only of planet formation processes, but also of presolar galactic, stellar, and interstellar processes. Credit: L. Nittler/IAC-UC3

Astronomers consider that a large fraction of interstellar dust in the universe is produced by the AGB stars. As their progenitors are of quite low mass, they evolve fairly slowly, not as slowly as stars with the mass of the sun, but they take gigayears of evolution time to reach the AGB phase. We know, as we will show below, that the age of the solar system is some 4.6 Gyr, which means that the universe was some 9 Gyr old, and the galaxies were some 8 Gyr old when the solar system formed. These 8 Gyr correspond, in broad terms, to the lifetimes of most of the stars which would have been in the AGB phase 4.6 Gyr ago, so the timing of the formation of our planetary system makes sense in a framework of understanding the evolution of the stars in our Galaxy. In 1987 the discovery of individual grains of silicon carbide and graphite with extreme isotopic anomalies in all measured elements was a breakthrough, showing that they were composed of matter from individual stars, and since then many types of presolar stardust grains have been identified in extraterrestrial material.

With this somewhat general scheme in mind, we can see how in a particular case the isotope analysis of meteoritic and cometary dust particles can be used to look for their formation sites.

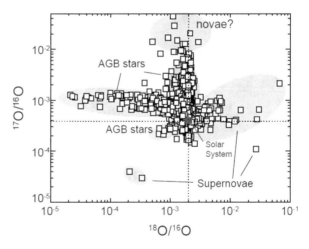

Fig. 11.2 Measurements of the ratios of the two less abundant oxygen isotopes ^{18}O and ^{17}O to the common isotope ^{16}O measured for meteorite grains identified as pre-solar by their generally anomalous isotope ratios. Pre-solar oxide and silicate grains are represented. The ratios of these isotopes in the solar system (Earth, Moon, Sun, and meteorites) are within the red square. Hypothetical sources based on the predictions for nucleosynthesis in different possible types of precursor objects are indicated in the light grey shaded areas. By far the greatest contribution comes from the AGB stars, However the oxide grains with low abundances of ^{18}O are apparently the result of nuclear processes not previously part of standard AGB models. Credit: L. Nittler

Figure 11.2 is a graph showing the ratios of three isotopes of oxygen, the common isotope, ^{16}O, and the two rarer isotopes ^{17}O, and ^{18}O measured in tiny (micrometre sized) meteorite grains which had been identified as pre-solar after determination of their anomalous isotopic composition using a range of elements. A great deal of work has been done on the physics of element production under different conditions within stellar interiors, and on the mechanisms for dispersing the products of this nucleosynthesis into space from the different types of stars. The broad predictions relevant to the production of the oxygen isotopes are shown in Fig. 11.2 as the grey shaded areas, each labelled with its corresponding stellar source. As could be envisaged the AGB stars are the principal sources of the material in these pre-solar grains. Figure 11.3 is a similar plot but for the isotope ratios of two different elements, carbon and nitrogen measured in presolar silicon carbide grains from a number of different sources, and also silicon nitride. These isotopes

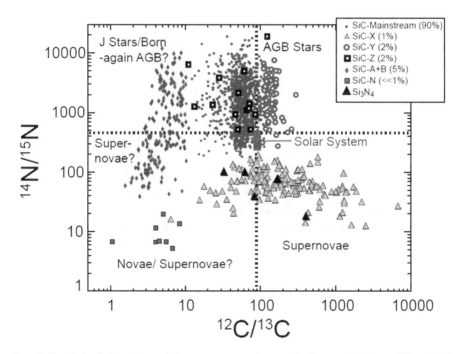

Fig. 11.3 Plot of the ratios of the common carbon and nitrogen isotopes ^{12}C and ^{14}N ratioed with their less abunant isotopes ^{13}C and ^{15}N respectively, measured in different meteoritic silicon carbide grains and silicon nitride grains. The range of solar system values is shown by the red rectangle. Different probable sources for the observed sets of data are shown by the light grey shaded areas. A wider range of probable sources than those shown in Fig. 11.2 is probed by this combination of isotopes as you can see. Credit: L. Nittler

probe a wider range of sources and you can see that in some grains the source appears to be supernovae rather than AGB stars.

Another, related way to study and identify presolar grains is the use of isotopes which indicate the decay of radioactive nuclei even those with lifetimes short relative to the age of the solar system. These latter are identified and measured by their known stable decay products. Steady advances in chemical and physical separation of components, in particular the development of the technique of secondary ion mass spectrometry, (and then nano-secondary ion mass spectroscopy) permitted submicron sized grains to be minutely analysed. These were found in AMM's, IDP's, in situ in meteorites and later in the dust resturned from comet Wild-2 which we will discuss below. Among the refractory grains which turned out to be presolar were those containing aluminium and magnesium oxides, and a variety of silicates.

11.3   Specific Samples of Extraterrestrial Material

11.3.1   Large Meteorites

11.3.1.1   The Murchison Meteorite

Although meteorites fall over the full surface of the Earth, most of the particles which reach the ground are very small. As we will see, these can be found and identified, and offer excellent research material, but the largest meteorites are of exceptional value because they can be collected almost immediately on impact, which avoids the effects of weathering on their physical and chemical properties. One the most studied meteorites fell in Australia, in 1969 near the town of Murchison, in the state of Victoria. A piece of the Murchison meteorite is shown in Fig. 11.4a, and a small grain is shown in Fig. 11.4b. It broke into numerous pieces, but the total mass collected from an area bigger than 13 square kilometres, was well over 100 kg.

The largest of the pieces has a mass of 7 kg. Research laboratories all over the world have taken fragments of this meteorite for analysis. The Murchison meteorite is a carbonaceous chondrite, which we have classified very briefly above. It has experienced alteration by aqueous liquids in the body on which it originated before arriving on Earth. It contains many calcium-aluminium rich inclusions (CAI's), millimetre-to-centimetre-sized ceramic-like objects which are the oldest objects in the solar system directly dated by radioisotope methods. They formed ~4.6 Gyr (billions of years) ago and this age is often

a

b

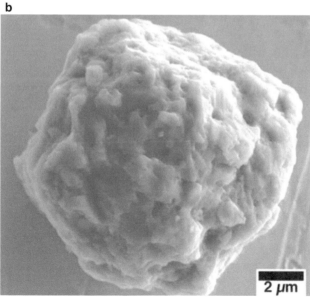

Fig. 11.4 (a) Piece of Murchison meteorite from National Museum of Natural History, Washington D.C. Carbonaceous chondrite, 22.13% iron, 12% water Credits: User:Basilicofresco, CC BY-SA 3.0 <https://creativecommons.org/licenses/by-sa/3.0>, via Wikimedia Commons. (b) A pre-solar silicon carbide grain from the Murchison meteorite, recently shown to be the oldest known material on Earth. Credit: Field Museum of Natural History, Chicago/Philipp Heck

taken as the starting point of the solar system. Murchison also contains a large variety of organic molecules, including more than 15 amino acids. There has been considerable discussion about whether there is any excess of left-handedness or right-handedness in these amino acids and in other organic compounds found in the meteorite. Laboratory analyses have shown slight excesses, but in general a rather balanced mix, which contrasts with the fact that most amino acids in terrestrial living organisms are left handed, and most sugars are right-handed. (The handedness refers to the rotation of the plane of polarisation of light passing through the material). In very general terms the products of this meteorite show that quite complex organic material can be delivered to the Earth via meteorites, and could have participated in the origin of the molecules of life. Although CAI's are the oldest dated solar system objects, in January 2020 analysis of some pre-solar silicon carbide particles in the Murchison meteorite showed that they must have been in interstellar space for 2.5 billion years before the Sun and planets formed, thus making them the oldest dated material found on Earth up to now.

11.3.1.2   The Allende Meteorite

It is curious that another even larger carbonaceous chondrite, even bigger than the Murchison meteorite, also fell in early 1969, but in Mexico, in the state of Chihuahua, near the village of Allende, a sketch map is included in Fig. 11.5.

The area within which the strewn fragments of the meteorite were found measures some 8 by 50 km. The landscape is a desert, which meant that the pieces were not quickly weathered before collection. Over two tons of specimens were collected in pieces ranging from grams to over 100 kg. This meteorite is probably the most studied in history, in part because it fell just before the first moon rocks were returned via the Apollo programme, and laboratories were gearing up for the analysis of this type of material. The meteorite is considered "stony", rather than "iron" or "stony iron". Like the Murchison meteorite it contains a dark matrix in which are embedded lighter chondrules, which are tiny stone spheroids found only in meteorites, and not in terrestrial rock. Larger inclusions, calcium-aluminium-rich inclusions, or CAI's, which we mentioned for the Murchison meteorite, were also found in Allende. The fraction of iron is 24% but, unlike Murchison, iron-nickel mix is virtually absent.

Figure 11.6 shows a piece of the Allende meteorite, giving an idea of its internal structure. The individual fragments of the meteorite have a hard

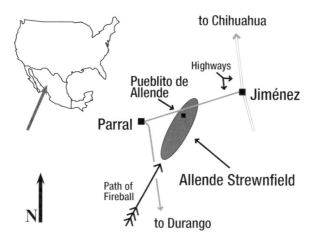

Fig. 11.5 Sketch map of the location of the Allende meteorite impact area. Credits: Milo44, CC BY-SA 3.0 <https://creativecommons.org/licenses/by-sa/3.0>, via Wikimedia Commons

glassy shell caused by the extreme frictional heating as they plunged through the Earth's atmosphere, so that analysis for age and history has to be performed on the interiors. The matrix and the chondrules have many different materials, predominantly the minerals olivine and pyroxene. Allende is rich in elements with high melting points such as calcium, aluminium and titanium, and poor in volatile elements such as sodium and potassium. There is evidence of radiation damage to the chondrules, which is also found in lunar basalt rock, but not in terrestrial equivalent rock, which has been protected from cosmic radiation by the Earth's atmosphere and magnetic field. There is no evidence of radiation damage to the matrix, which implies that the chondrules must bave been damaged before the matrix accumulated around them in the cold accretion process during the early formation of bodies in the solar system. Within Allende, and other meteorites, there are tiny specks of matter with anomalous ratios for the isotopes of a number of elements, including aluminium, calcium, barium and neodymium. Anomalous means, as we have seen, that the ratios are different from those found in the Solar System in general. Subsequent studies have shown anomalies of this type in the gases xenon, krypton, and nitrogen. These findings have been quantified to show that parts of chondrites must have been formed earlier than the Solar System itself, from materials ejected from different types of stars, as we saw above. In addition to the inorganic material there are small quantities of organic molecules in Allende, including amino acids some of which are not known on

Fig. 11.6 A chunk of the Allende meteorite, showing its internal structure. The comparison cube has side length 1 cm cm Credits: The original uploader was Bennoro at English Wikipedia., CC BY 3.0 <https://creativecommons.org/licenses/by/3.0>, via Wikimedia Commons

Earth, and an extraterrestrial protein, containing iron and lithium was tentatively identified early in 2020, although this is controversial.

11.4   Material from Comets

11.4.1   Material Brought to Earth from a Comet: Comet Wild 2

On February 7th 1999 NASA launched the fully automatic Stardust space probe whose main aim was to collect a sample of cometary dust from the long period comet Wild 2. After several manoeuvres and after passing close to asteroid Anne Frank in 2002, Stardust made a close flyby of the head of comet Wild 2 in 2004, before returning to Earth in 2006 to release a Sample Return Capsule, which landed on the Bonneville Salt Flats in Utah. With a further manoeuvre the spacecraft then flew on to a rendezvous with Comet Tempel in 2011. The encounter with comet Wild 2 ocurred with a separation in distance between the spacecraft and the comet's solid head of 237 km and a relative velocity of 6 km/s. The probe was on the sunward side of the comet, which overtook it. During the close approach the probe deployed a specially designed sample collection plate, shown in Fig. 11.7a.

The plate was made of "aerogel" an extremely low density silica glass such that its ratio of surface area to mass was over 1000 times that of a solid block of the same material. During the approach the probe was within what is normally called the coma of the comet, where material emitted from the surface of the comet's nucleus due to the enhanced solar radiation and particle flux is high. The sample plate collected a useful quantity of this material, and a cross-section showing the dust tracks within it is shown in Fig. 11.7b The plate was returned to Earth 2 years later when Stardust made its pass above the atmosphere. The descent of the return capsule through the atmosphere was dramatic. It entered at 12.9 km/s the fastest reentry of any object fabricated by people. Its velocity was reduced from Mach 36 to subsonic speed in 110 seconds, with a peak deceleration of 34 g! It reached a temperature of 2900 C, before parachuting to the ground on a US Army range in Utah, from where it was transported to the Johnson Space Center in Houston for initial analysis. This was performed in a clean room with 100 times fewer particles per cubic metre than a hospital operating theatre to minimise terrestrial dust contamination. Over one million dust specks were estimated to be in the aerogel, of which more than ten had sizes greater than 0.1 mm, and the largest had a size of ~1 mm. The dust was analysed by a professional team and also by volunteers via an online "citizen science" project Stardust@Home. In the first set of papers with results from the mission, a range of organic compounds was found, including two which contain nitrogen in a state usable in biological processes. Carbon chain molecules with longer chains than those observed in the diffuse interstellar medium, silicates, both amorphous and crystalline, (olivine and pyroxene), were found but no hydrated silicates or carbonates, implying no water processing in the comet, but in subsequent research iron and copper sulphides were found, implying the presence of some liquid water. In 2009 NASA announced that the amino acid glycine had been found in the Comet Wild 2 sample, the first time this basic building block of DNA had been found in a comet, although it had previously been identified in meteorites. This finding strengthens the general idea that life components may well be generally distributed in the universe. Finally in 2014 the discovery of particles of unaltered interstellar dust from the Stardust sample was announced.

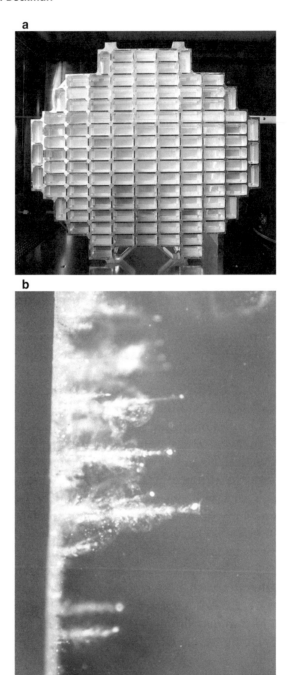

Fig. 11.7 (a) "Aerogel" collector plate used in the Stardust project to collect dust particles from Comet Wild2 when the Stardust probe made its closest pass. The material of the plate is a specially "foamed" glass, which has 1000 times more area for its mass than a normal glass plate. (b) A section of the airgel plate from the Stardust mission showing tracks of dust particles from Comet Wild 2. Credits: NASA/JPL

11.4.2 Other Missions to Comets

As this chapter is discussing specifically material which can be, and has been, examined directly in laboratories, and is thus fully within the scope of experimental science, I will go into very little detail about missions by space probes to other comets, even though they are clearly important and have yielded very interesting information.

Giotto: Halley's comet flyby.

The first mission to make a really close approach to a comet was ESA's Giotto probe. Giotto was in fact the tip of the lance of a combined operation in which two Japanese probes and an existing US probe made measurements from a fairly long distance, two Russian probes, Vega 1 and Vega 2 helped to locate the nucleus within the coma, and with the information from these Giotto could target the nucleus with precision. The Vega probes flew past the comet, at just under 10,000 km from the nucleus, on March 6th and March 9th 1986, respectively, and on March 14th Giotto flew by at 586 km distance. The ESA Giotto team doubted that it would survive, because of impacts by the stream of small bodies emanating from the comet's surface as it was closing in towards the Sun. Giotto was in fact impacted and temporarily lost its orientation, but the engineers recovered it in time to take photos with a multicolour camera. The camera was destroyed by another impact shortly after closest approach. Figure 11.8 is a picture of the comet nucleus taken with the camera, showing that the it was peanut shaped, which judging from the photos I have found to illustrate this chapter, seems to be the norm for comet nuclei. It also shows the material spraying out of the sunward side of the comet as it approached the Sun. It was some 15 km long by between 7 and 10 km broad, and porous, with a density around half that of water. One of the most surprising results of the probe was the low albedo of the comet, it reflected only some 2% of the sunlight falling on it, because its dusty covering was largely black.

The plasma and ion mass spectrometer on board Giotto showed that this dust was carbon rich, consistent with this low albedo. At the time of the fly-by the rate of mass loss from the nucleus was measured at some three tons per second, from the spurting jets on its sunward side. of which 80% was water, 10% carbon monoxide, 2.5% a mixture of methane and ammonia, with traces of other hydrocarbons, iron, and sodium. The ratios of carbon, nitrogen and oxygen abundances were similar to those of the Sun, an early indication that comets are a part of the nebula from which the Sun was formed. The solid particles ranged from the sizes found in terrestrial smoke to masses of order a

Fig. 11.8 Photo of the nucleus of Halley's comet taken from ESA's Giotto spacecraft at closest approach during 1986. The rate of mass loss from the comet was estimated at three tons per second, of which 80% was water, when this picture was taken. Credits: ESA/MPAe

few hundred milligrams. An on-board mass spectrograph allowed a degree of analysis of their contents. They were of two distinct types, one containing carbon, hydrogen, nitrogen and oxygen, the other with calcium, iron, magnesium, iron and silicon. Another interesting finding was that the interaction of the comet's atmosphere with the solar wind gave rise to a zone with zero magnetic field close to the leading surface of the comet. Giotto also detected the bow shock where the solar wind met the comet's outflowing atmosphere. As a spaceflight achievement the Giotto probe accomplished several "firsts" as well as being the first to achieve a close flyby of a comet nucleus. In fact it flew close to two comet nuclei, because after its rendezvous with Halley, the ESA engineers brought it to an encounter with comet Grigg-Skjellerup in July 1992, with a fly-by at less than 200 km distance. Giotto's instruments detected dust particle outflow from the comet, and the bow shock due to its interaction with the solar wind. En route to Grigg-Skjellerup it had a gravity assisted fly-by of the Earth, which was the first time that this maneouvre had been accomplished, and it was repeated again in 1999. Giotto had been put into hibernation after its encounter with Halley, and was revived when it reached Grigg.Skjellerup, the first time that this had been accomplished on a space mission.

Rosetta: The first probe to orbit a comet.

ESA's Rosetta project was a first in space engineering, the first probe to orbit a comet. It was launched in March 2004 and reached its goal, comet 67P Churyumov-Gerasimenko on 6th August 2014 after a very carefully tailored set of orbits, involving flybys of the Earth, Mars, and asteroids Lutetia and Steins. It went into an orbit taking it between 30 and 10 km from the comet, and began to take a long sequence of images. One of these is shown in Fig. 11.9a where you can see water vapour and CO_2 spurting out of the surface of the comet's rocky nucleus.

On 12th November its module Philae made a landing on the comet, or rather it made three landings, because its anchoring mechanism failed to operate, and Philae bounced off twice before falling back to make an awkward final landing. The rendezvous ocurred between the orbits of Mars and Jupiter, and Rosetta stayed in orbit around the comet until after it had reached perihelion, so contributing to our knowledge of cometary behaviour as the Sun was approached. Rosetta carried a suite of 11 instruments based on electromagnetic radiation, from optical, infrared and ultraviolet spectrographs to microwave and radar sounding probes. It also carried four instruments to measure the properties of the gas and dust particles *in situ*. The lander Philae carried ten additional instruments, which included a drilling facility to transfer matter to its analytic instruments for measurement. My aim here is only to summarise a few key results. One of the first results noted by Rosetta was that the magnetic field around the comet oscillated at 30–50 millihertz, but Philae did not detect a magnetic field on the solid nucleus, so the effect is attributed to the solar wind. Probably the most striking result was that the ratio of

Fig. 11.9 (a) Comet 67P Churyumov-Gerasimenko spurting water vapour and CO2 due to the effects of solar radiation and the solar wind as it approached the Sun. Photo taken from Rosetta as it orbited the comet's nucleus. (b) Mosaic of a portion of the surface of Comet Churyumov-Gerasimenko taken from Rosetta. Credits: ESA/Rosetta/ NAVCAM, CC BY-SA IGO 3.0

deuterium to hydrogen in the water vapour released by the comet was three times greater than that of terrestrial water. One of the previously well regarded scenarios for the supply of water to the nascent Earth was that it came from comets. This measurement by Rosetta "put a damper" on that idea. The release of water vapour from the comet increased by a factor 10 between June and August 2014 as the comet moved towards the Sun. Organic non-volatile compounds were found everywhere on the surface of the comet, and in the dust in its atmosphere, in the form of polyaromatic solids mixed with sulphides and iron-nickel alloys. These large molecules are similar to those found in carbonaceous chondrites, but no hydrated minerals were found, in contrast to the chondrites. The only amino acid found was glycine, but there were many less complex organics, such as acetamide, acetone, and methyl isocyanate. A big surprise was the detection of large proportions of free molecular oxygen in the atmosphere of the comet. There have been almost 5000 publications about the results from the Rosetta mission, between professional journals and conference proceedings, so any attempt to summarise them would require a book in its own right.

11.5 Missions to Retrieve Material from Asteroids

11.5.1 Material Brought to Earth from an Asteroid Itokawa

In 2003 the Japanese Space Agency (JAXA) launched the Hayabusa space probe to explore the near-Earth asteroid Itokawa. It made its rendezvous with Itokawa in September 2005. After hovering in a heliocentric orbit near the asteroid and studying its shape, spin, topography and colour, it landed twice in November and collected a sample of tiny grains before returning with them to Earth on June 13th 2010. An image of Itokawa taken from Hayabusa is shown in Fig. 11.10.

The Hayabusa mission was plagued with problems, and it is remarkable that it managed to achieve what it did. One of its main aims was to test a set of microwave discharge ion engines, the first time this kind of propulsion had been used in a space mission. Although they gave many problems, with partial failures and re-ignitions, the overall programme was a success; several of them were kept going during the full trajectories of the outward and return journeys. Then the probe was not designed for a complete touchdown, but it did touch down briefly. It was supposed to fire tiny projectiles at the asteroid to release dust, but in practice the dust collected was thrown up by the probe itself as it touched down. An auxiliary probe, MINERVA was supposed to land, but failed to do so. Nevertheless Hayabusa delivered the sample capsule to Earth, where it landed at Woomera in Australia after falling dramatically through the atmosphere in a similar manner to the Stardust sample return

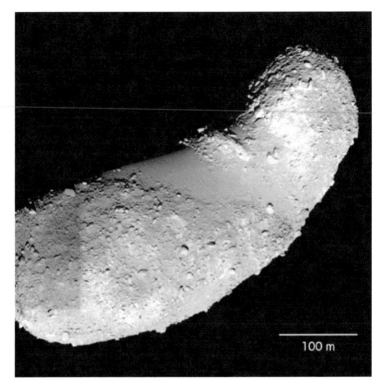

Fig. 11.10 Asteroid Itokawa, photographed from the Hayabusa space probe. Credit: JAXA

capsule from Comet Wild 2. A photograph of the capsule as found when it landed is shown in Fig. 11.11.

In 2011 the first scientific results from Itokawa were published, based on the highly refined technique of nanoscale secondary ion mass spectroscopy, which has become standard in meteorite research. The dominant components of the grains are olivine and pyroxene, making up some 80% of their mass. A key result was that the oxygen in the samples was depleted in the ^{16}O isotope compared with terrestrial rock, but very similar to the proportions found in chondrites from meteorites which have fallen to Earth. This places Itokawa in the S-class of asteroids, with predominantly silicate composition, as observed from the ground by spectroscopy. In 2018 Japanese scientists published a study of radioactive isotopes found in Itokawa, measuring uranium isotopes and the lead into which they have decayed. Combining isotopes in the series ^{238}U- ^{206}Pb, which has a decay half-life of 4.47 Gyr, with isotopes in the series ^{235}U- ^{207}Pb, which has a decay half-life of 700 Myr, they found an age for the crystallisation of the material of Itokawa of 4.64 (+0.18) Gyr (billions of

Fig. 11.11   The sample return capsule from the Hayabusa asteroid probe, just after it landed at the Woomera site in Australia. Credit: JAXA

years) which is close to the best age estimates for the formation of the Earth, just a little earlier. But they also detected a shock occurring 1.5 Gyr ago, which they attribute to an impact which split Itokawa from a larger parent asteroid. Then in 2019 another research group found water in the supposedly anhydrous grains from Itokawa, with a ratio of deuterium to hydrogen very similar to that found on Earth, close to 1.5 parts in ten thousand. They estimate that the water fraction on the asteroid is a few hundred parts per million. Their conclusion from this is that asteroids of this type could have been responsible for bringing water to the Earth during and after its main formation period.

11.5.2   Hayabusa 2: Asteroid Ryugu

Spurred on by their success with Hayabusa, the Japanese space agency prepared an asteroid lander, Hayabusa 2, which was launched towards comet Ryugu in December 2014, and reached its objective in June 2018. On 21st September it succeeded in landing two small rovers on Ryugu, and on 22nd February 2019 the Hayabusa 2 spacecraft itself landed successfully but briefly on the asteroid to collect a sample. A little later a small explosive charge was released from the spacecraft while still in orbit, which blew a 10 metre diameter crater in the asteroid. In this crater the spacecraft landed again, and collected material which had been well below the surface. In this way a comparison between space weathered material and shielded material will be possible. Hayabusa 2 remained on the surface of Ryugu until December 2019, and began its return to Earth. The capsule with the asteroid samples was dropped into the atmosphere on December 6th 2020, landed successfully at the Woomera Test Range in Australia and was transported to Japan for laboratory studies.

11.5.3   OSIRIS-REx: Asteroid Bennu

On 8th September 2016 NASA launched its asteroid probe OSIRIS-REx towards the asteroid Bennu. After passing close to the Earth twice for gravity asssits the probe reached its target on december 3rd 2018 and braked to a relative approach velocity of only 20 cm/s. It has been in orbit since then, with a set of instruments surveying and photographing the asteroid, and has detected a small amount of activity, the release of dust particles from the surface, which had not been expected. I am writing this 2 days after, on October 20th 2020 OSIRIS-REx descended successfully and briefly extended a special probe to collect a sample of small rocks, up to a couple of cm diameter, from the surface of Bennu. Figure 11.11 is an image taken by a camera on board OSIRIS-REx of the extended sample probe in contact with the surface. The maneuvre acronym was TAG, for "Touch and Go", which describes not only the manner of collecting the sample, but also the difficulty of operating correctly in the extremely low gravitational field of the asteroid. OSIRIS-REx should return to Earth and send down a capsule with samples to land in the Utah desert on 24th September 2023 (Fig. 11.12).

11.6   Rocks from the Moon

One of the objectives of putting human beings on the Moon was to study its physical properties and history, and one of the main routes for this study was the collection and recovery of a representative sample of lunar surface material. There were six lunar landings during the Apollo project, Fig. 11.13 shows where these took place, and Fig. 11.14 shows Astronaut Harrison Schmitt, a professional geologist by training, collecting a sample during the Apollo 17 mission.

Fig. 11.12 The sample collecting probe from OSIRIS-REx photographed during its brief contact with the surface of asteroid Bennu. Credits: NASA-GSFC/ University of Arizona

One of the criteria for the selection of the landing sites was to explore a variety of selenological features, and to bring back a selection of lunar rock representative of these features. During the decade of the 1970s lunar science flourished as never before, or since, stimulated by the results obtained in laboratories around the world from lunar material. Here I will give only some examples to show what kind of work is done in this field. We can start with classical work by Grenville Turner of the University of Sheffield starting in 1970, using radioactive dating. He used the method of estimating the fraction of radioactive potassium ^{40}K which had decayed into the ^{40}Ar isotope of argon in a sample of regolith from a number of different lunar sites. Before the measurement can be made, the sample has to be taken to a suitable laboratory nuclear reactor, and irradiated with neutrons, which convert the ^{39}K isotope of potassium into the ^{39}Ar isotope of argon. The argon, with its mixture of isotopes, is then made to diffuse out of the sample by heating, and the ratio of the two isotopes is measured using a mass spectrometer. This allows the current ratio of ^{40}K to ^{40}Ar to be measured, and this ratio gives the age of the sample, because the decay rate of ^{40}K to ^{40}Ar is accurately known from standard terrestrial laboratory experiments. In order to obtain reliable information about the sample age it is necessary to assume that no argon has escaped, either on the Moon itself, or after collection and transport to Earth. The way to handle this problem is to collect the gas which is emitted during the

Fig. 11.13 Six Apollo missions successfully landed on and departed from the Moon between July 1969 and December 1972. Top, clockwise: James Irwin salutes the flag at Hadley Rill; Harrison Schmitt collects rock samples in the Taurus-Littrow Valley; Buzz Aldrin's footprint in the lunar regolith; Charlie Duke placed a photo of his family on the Moon and took a picture of it; Edgar Mitchell photographs the desolate landscape of the Fra Mauro highlands; and Pete Conrad jiggles the Surveyor 3 probe to see how firmly it's situated. Credits: Bob King/Sky & Telescope/NASA

heating process in successive stages as the temperature is raised. The idea is that the argon which comes off first is that closest to the surface of the sample and for which there is the highest probability of argon escape, while the last emitted argon is that most bound to the rock. Successive age estimates of the gas should show increasing values, and the last released gas, or preferably the last few measurements of released gas, should approach the true age asymptotically. This proposal was shown experimentally to work well, and for seven rock samples from Apollo 11 taken from the landing site on the Mare

Fig. 11.14   Astronaut Harrison Schmitt collecting moon rocks during Apollo 17 mission
Credit: NASA

Tranquilitatis, ages between 3.52 and 3.92 Gyrs were obtained, indicating that this Mare was formed during that time period. Subsequent determinations using the radioactive decay of rubidium to strontium gave a very similar result. But it was also possible to show specific loss of ^{40}Ar at given points in time, presumably due to specific local meteor impacts, with one recent impact timed within the last 100 million years. In work with the same samples Turner measured the isotope ^{38}Ar which is produced by the interaction of cosmic ray particles with potassium, chlorine and calcium, and by finding the ratio of ^{36}Ar to ^{38}Ar he could estimate for how long the rocks in the sample had been exposed to the cosmic rays, persumably from the solar wind. His estimates of these "cosmic ray exposure ages" for different rocks were in the range of a few hundred million years, giving the times when these rocks were brought to the surface of the Moon. The same method of dating using the decay half life of radiactive potassium ^{40}K to the argon isotope ^{40}Ar gave an age of 3.33 Gyr for a basalt sample collected by the Apollo 15 astronauts from Halley Rill, at the edge of the Mare Imbrium. But these basalt flows were shown to have occurred at least 500 million years later than those of the Mare Imbrium itself, as measured using rocks from the Fra Mauro formation brought back by Apollo 14. An implication is that the Mare was not formed following an external impact, but was due to intrinsic volcanic activity on the Moon in its early epochs. This general picture was strengthened by the measurement of magnetism in a significant number of returned samples from various parts of the Moon. A very small part of the magnetism in a given sample was removable by demagnetisation procedures, but the major part was not. Researchers concluded that the variable part had been artificially induced during collection and tranport, but the contant part was a "frozen in" property of the rocks. Comparing the positions of origin a picture emerges of a uniformly organised directional magnetic field, operating between 3.8 Gyr ago and 3 Gyr ago. This is consistent with a liquid core for the Moon during that epoch, and with the explanation of the Maria as extensive lava flows from this liquid core.

The dating of the rock samples gave another method of understanding the Moon's surface, using the multitude of craters caused by meteorite impacts. As the only weathering on the surface is by the solar wind, craters larger than a few metres in size will remain visible for many millions of years. By 1972 it was possible to estimate the rate of crater formation as a function of time using sample dating, with the overall conclusion that before 4.1 Gyr ago the meteor impact rate on the Moon was a thousand times greater than it is today. The rate declined steeply between then and just over 3 Gyr ago, and has remained virtually constant since then. The latter conclusion was reached by counting the numbers of craters as a function of size within the huge Tycho crater, with its own age determined as 109 million years, on the floor of the Copernicus crater, which has an age of 800 million years and around the Apollo 12 landing site, with its estimated surface rock age of 3.26 Gyr. Calibration using these data has allowed researchers to use crater counts as a measure of the age across the surface of the moon. As complementary information the Apollo 16 mission provided samples of lunar regolith in the highland areas, from which ages could be determined on the basis of the decay of uranium isotopes into lead, giving an average age of 4 Gyr.

11.6.1   Are the Tektites of Lunar Origin?

One question to which the Apollo moon rock samples gave a definitive answer was the origin of the tektites. These are rocks whose glassy texture shows that they are the result of a major impact, either by a meteorite or even by the head of a comet. There are four major fields where tektites are strewn, in North America, the Czech republic, the Ivory Coast, and a very wide field covering south Asia and Australia. Before the arrival of the moon rock samples there was much debate about whether they were caused by impacts on the Earth or on the Moon. The chemical composition of the tektites is very different from that of the lunar maria basaltic rocks, and in particular the latter have chromium contents two orders of magnitude greater than that of the tektites, and there are many other discrepancies. The rocks from the lunar highlands have far more aluminium and far less silicon dioxide than have the tektites. The ages of the lunar maria (3.2–3.8 Gyr) and of the lunar uplands (>4 Gyr) are an order of magnitude greater than the material of the tektites in South Asia and Australia as measured by the radioactive decay of rubidium to strontium, and the lead in the lunar rocks is mostly from radioactive decay of heavier elements, whereas the tektites have lead of terrestrial isotopic composition. So one clear result from the Apollo mission is that tektites are not made of lunar material.

11.7 The Genesis Mission: Oxygen Isotopes in the Solar Wind

In August 2001 NASA launched Genesis, a probe to measure the composition of the solar wind. The aims were to sample the wind in situ and also to collect samples and return them to Earth for analysis. It was the first NASA mission to return samples since the Apollo Moon landings nearly 30 years previously. On September 8th 2004, after 3 years sampling in space the Genesis payload was returned to Earth, but its landing in Utah was marred by the failure of its parachute to open. The crash landing ruptured the capsule and jeoardised the results, and the majority of the detectors which had been exposed to the solar wind in space had to be discarded. Nevertheless it was possible to remove the surface layers of certain types of detectors, and to check the detected solar wind particles in their interior zones with extreme care for contamination. Useful results were obtained about key solar isotope ratios, among the most important those for the isotopes of oxygen. It is not possible to separate spectral lines due to the three isotopes of oxygen, ^{16}O, ^{17}O, and ^{18}O using even the highest resolution spectroscopy from the ground but it was possible to measure these ratios directly using the Genesis data. Compared to the standard used in terrestrial geology, which is based on sea water, the oxygen measured with Genesis was enriched in ^{16}O; the ratio of ^{16}O:^{17}O was 10% higher than in seawater. This is similar to measurements of the same ratio made on the Apollo surface samples, known to be exposed to the solar wind. It is interesting that the values measured in widely scattered chondrites on Earth including those found in the Murchison meteorite, which are attributed to values present in the nebula which gave birth to the Solar System, show an increment over sea water values in the range 23 to 25 parts per mil.. However values of these isotope ratios in the Martian meteorites found on

Earth, on the Moon in the Apollo samples from within rocks protected from the solar wind, and similarly on Mars as sampled by the instruments on the Mars landing vehicles, are all similar to each other, but different from those of the solar wind, and of the oldest meteorites. The clear difference between the solar and solar nebula oxygen isotope ratios and those of the Earth, Moon, and Mars is shown in Fig. 11.15.

This finding, which has gradually become clearer during the first two decades of the present century, is certainly a powerful clue about how, and when, the inner planets formed from the original solar nebula. There are specific scenarios which have been proposed to explain the oxygen isotope dichotomy. One of them is that the material in the protoplanetary disc was augmented by a nearby type II supernova, a hypothesis which was first proposed to explain the presence of measurable amounts of the radioactive isotopes ^{28}Al and ^{41}Ca in meteorites, isotopes which are too short lived to have remained in detectable quantities if they had been in the original solar nebula without a later injection. But an alternative explanation is that the dichotomy

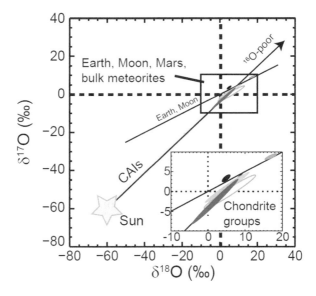

Fig. 11.15 Ratios of oxygen isotopes found in rocks on the Earth, Moon, and Mars, (dashed line) compared with the same ratios found in the refractory inclusions in chondrites and the solar wind measured by the Genesis detectors, The distinct differences in behaviour show that the inner planets were formed from material which received an additional set of elements from a different source after the Sun formed. The general idea has been that the original solar nebula had dust from AGB stars, and that the planets had an additional contribution from a nearby type II supernova, but more recently photochemical effects in the placental molecular cloud or the preplanetary accretion disc have been a preferred explanation. Credit: L. Nittler

reflects a large-scale photochemical effect either in the disc phase, or in the parental molecular cloud. Specifically the general idea is that ultraviolet light can dissociate the CO molecules which are abundant in molecular clouds, and presumably in pre-planetary discs. This dissociation requires UV of a slightly different wavelength depending on whether the CO contains one or other of the oxygen isotopes ^{16}O, ^{17}O, or ^{18}O. Because ^{16}O is so much more abundant, at points within the cloud beyond a certain distance from the illuminating star all the photons which would dissociate $C^{16}O$ will have been absorbed, but the $C^{17}O$ and the $C^{18}O$ are still dissociated so the ^{17}O and ^{18}O will be released and can react. When they react with hydrogen this will produce water enriched in these heavier oxygen isotopes, and will eventually produce enriched silicates. So in the a preplanetary cloud there would be two zones, one which has its ^{16}O enriched compared to the Earth, and another with higher proportions of ^{17}O and ^{18}O in its silicates. The solid planetary material we measure was produced from a mixture of material from these two types of zones. This example gives just a taste of the effort and complexity which modern physical techniques are bringing to Solar System studes based on the acquisiton and direct measurement of samples from outside the Earth.

11.8   Martian Meteorites on Earth

Since robotic lander vehicles have set down on Mars it has been possible to make in situ analysis of the chemical content and history of Martian rocks. This has led to a particularly interesting corollary: certain meteorites with specific compositions not repeated either on Earth or in other meteorites, have been identified as emanating from Mars. The scenario is of a powerful impact of a meteorite onto the Martian surface releasing rocks from the planet, which could escape due to the relatively low gravity. A small fraction of this material encountered the Earth and fell as meteoritic material. In the 1980s it became clear that three groups of meteorites, referred to as the SNC groups (Shergottites, Nakhalites, and Chassignites, named for the places where their first members were found), which had already been shown to have compositions different from other types, were very likely from Mars, following rock analysis by the Viking landers in 1976. In 1983 trapped gases in a Shergottite were found to be similar in composition to those in the Martian atmosphere analysed by Viking. In 2000 an article by Treiman, Gleason and Bogard surveyed the arguments and concluded that there was a great probability that the SNC meteorites are from Mars. They have isotope ratios consistent among themselves and different from those of other meteorites, and from terrestrial material. Particularly well known Martian meteorites are the Nakhalites, named after Nakhala in Egypt where in 1911 a meteorite weighing 10 kg in total fell and was split into at least 40 fragments.

Figure 11.16 shows a Nakhalite which was found in the antarctic The chemical and petrological composition is sufficiently specific that 21 meteorite finds in different places, from Antarctica (7), the north of Africa (11), and the Americas (3), with sizes ranging from nearly 14 kilograms down to 22 grams have been classifies as Nakhalites. The basic mineral composition is 80% pyroxine and 10% olivine, with the remaining 10% a matrix or mesotastis, and the larger pieces are covered with black glassy crust produced by heating during passage throught the Earth's atmosphere. The identification of their origins on Mars is due to a combination of analyses of chemical, isotopic and chronological data. They were formed from magma some 1.3 Gyr ago, and were ejected from Mars some 10.75 Myr ago by an asteroid. Making the transit from a Martian orbit to a terrestrial orbit has taken almost all of these 10.75 million years, and they have all fallen to the ground within the past 10,000 years. Their crystallisation ages compared with crater count chronology for Mars suggests that they were formed in large volcanic features such as Syrtis Major and detailed analysis shows that they were suffused with liquid water 620 Myr ago, which disolved some of the olivine and deposited salt minerals in its place. In this way the nakhalite meteorites have shown the

Fig. 11.16 Nakhalite meteorite weighing a little over 1 kg found in the Antarctic at 750 km from the south pole. Isotopic analysis has proved that the Nakhalites were thrown off the surface of Mars by a meteorite impact, and took some 10 million years to reach the Earth (cube of 1 cm side shown for size comparison). Credits: National Museum of Natural Science/El Pais, Madrid

undoubted geological presence of major quantities of liquid water on Mars. But the big question then arose about whether the availability of liquid water, implying temperatures where living cells could survive, also implied that there had been life on Mars at some stage, even though now the conditions are adverse. One particular meteorite found in the Allan Hills of Antarctica, ALH 84001 has come under particular scrutiny since polycyclic aromatic hydrocarbons were identified in it, at levels increasing away from its surface, while contamination on Earth would be greatest at the surface. Other Antarctic meteorites do not contain PAH's. (but PAH's are very common in cool clouds in the interstellar medium). One group of researchers identified small ovoid and tubular structures which they thought might be nanobacteria fossils, but a world expert in terrestrial bacteria countered with the opinion that the structures look nothing like bacteria to him. Still pushing for evidence a different group claimed that the presence of magnetite in the meteorite could also be evidence of biological activity, but the majority view is that magnetite crystals can well be grown in situ by a purely physical-chemical process. As this chapter was being written a report was published of the discovery of organic compounds in ALH 84001 containing nitrogen. This is certainly a small step in the direction of life molecules, but although direct evidence from the Discovery surface rover for the historical presence of liquid water on Mars has been found, real evidence for life is at present considered weak or absent (nevertheless we should be mindful of the statement that absence of evidence is not evidence of absence).

11.9   In Situ Measurements of Martian Composition

There has been a steady stream of unmanned probes sent to Mars, the majority by NASA, some by ESA, and some by the Russian space agency. Mars has fascinated both laymen and scientists because it is judged to be the most likely place in the Solar System outside the Earth where life could form. One of the keys to whether a planet in this, or any other, system of planets is habitable is the presence of liquid water. The emotional point about this is that when we consider the probability that life exists elsewhere in the universe it would be critical to know whether life has developed independently on more than one planet within the Solar System. If we know this, our general knowledge of how life has developed on our own planet will allow us to plan our theory and experiments about extraterrestrial life in general. This was the driving element of what has been a long and successful scientific exploration programme, above all by NASA, of our neighbouring red planet. And the search for liquid water has been one of the basic aims. But as the story and the results of Martian studies are so extensive, and as techically the only samples of Mars that we have in our laboratories are the Martian meteorites, I limit myself to the briefest selection of the results which I have chosen simply because they interest me.

The first probes to land successfully on Mars were the NASA's Vikings, 1 and 2. Both landed in 1975, and Viking 1 returned signals until 1982, whereas Viking 2's signal was lost in 1980. The next NASA lander was not until 1996, when Mars Pathfinder survived only 9 months sending back information. The lengthiest and most productive programme has been the Mars Exploration Rover (MER) series, MER-A, "Spirit Rover" which landed in 2004 and moved around making various experimental tests until its signal was lost in 2010, MER-B, "Opportunity Rover" which landed in 2004 and worked industriously until 2018, and MER-C "Curiosity Rover" which landed in 2012 and is still operational at the time of writing. Figure 11.17 is a "selfie" by the Curiosity Rover, showing it in action.

A curiosity about its activity is that since 2017 it has been programmed to select its analysis targets using artificial intelligence to optimise the selection

Fig. 11.17 Curiosity Rover aiming a laser at a Martian rock; the products can then be analysed either spectroscopically or mass spectroscopically by the instruments carried on the Rover. Credits: NASA/JPL/Caltech

without constant uplinks from Earth. In addition NASA sent the Insight Mars lander to study Martian seismology which landed in May 2018 but lasted only until October 2019. The landing sites were chosen to sample a range of Martian terrain, and one of the most comprehensive results was that the composition of the rocks varies very little over the whole planet. Olivine, for example, is very widely spread. In the presence of liquid water this would weather into clays, so there cannot have been major amounts of liquid water on the surface of Mars in the past few million years. Although there are differentiated plains and highlands, the high winds blowing frequently across the surface of the planet have mixed the surface composition. There is ubiquitous fine dust on the surface, and the three rovers found it to be magnetic, based on the mineral magnetite, and also on titanium content. The magnetisation implies an aligned magnetic field, which in turn implies that Mars has a liquid core. The dark areas of Mars are basaltic, comprising olivine, pyroxine, and plagioclase feldspar, which is similar to the composition of the Lunar maria, and the Earth's ocean floor. Sedimentary rocks are common, and they show mixed action of air, vulcanism, and water in early epochs of the planet. The most abundant carbonates, calcium and magnesium carbonate, found in places in up to 30% by weight, are evidence of high concentrations of carbon dioxide in the past, and historic clay minerals show that liquid water was present in major quantities in past epochs, hundreds of millions of years ago. In one particular experiment by the Spirit rover, the mineral goethite was found in the Columbia Hills formation, by the technique of Mössbauer spectroscopy, a result which also shows the presence of liquid water in past epochs of the planet. And for those looking keenly for an environment which could favour life, in the first drilled sample of an in situ Martian rock, the John Klein rock at Yellowknife Bay in Gale crater, Curiosity rover detected the presence, in the bound rock, of water, carbon dioxide, sulphur dioxide hydrogen sulphide, chloromethane, and dichloromethane. A very general result about the Martian atmosphere is that the isotope ratio ^{13}C to ^{12}C in Martian carbon dioxide, (which is the dominant molecule in the atmosphere) of 45 parts per mil is indicative of considerable escape of the atmosphere, because ^{13}C is heavier than ^{13}C which makes it just a little harder for $^{13}CO_2$ to escape. We know that Mars has managed to retain a thin atmosphere, in spite of having a surface gravity only 38% of the Earth's because its surface is considerably cooler, so the velocities of the gas molecules are lower than in the Earth's atmosphere, which limits the probability of a molecule's reaching escape velocity. One of the most interesting results of the analysis of the Martian atmosphere is that the Argon isotopic ratio $^{40}Ar/^{36}Ar$ is the same as that found in the Martian meteorites found on Earth, described earlier in this chapter. This measurement, by the Curiosity rover, is the firm confirmation that Mars is the origin of these meteorites. But there was a previous indicative confirmation by the Spirit rover which analysed rocks found on the Martian plains, showing them to be of very similar composition to the Shergottite family of terrestrial Martian meteorites. An unexplained event in the Martian atmosphere was a sudden "spike" of methane concentration in the atmosphere on 16th December 2014, detected by the Curiosity rover, which rose to 7 parts per thousand million, during a single day, then fell back to its general level of around one tenth of this value. It is almost certain that this was a phenomenon local in space as well as in time, but no specific explanation has been offered. And finally in this rather ad hoc collection of nuggets from in situ Martian exploration, meteorites have been found strewn over the Martian surface, including "heatshield rock" close to the discarded heatshield of Opportunity rover, the first meteorite found on another planet, composed of 93% iron and 7% nickel. Perseverance Rover, with its tiny helicopter drone Ingenuity, landed on Mars on February 18th 2021. Watch out for its results!

This chapter has dealt with messengers which may not be as exciting for some physicists as gravitational waves, neutrinos, or even cosmic rays, but solid and gas infused matter arriving directly to Earth, or brought back from neighbouring bodies in the solar system is susceptible to the most minucious modern physical and chemical techniques of analysis, and is helping us to consolidate our knowledge and understanding of the universe. The reach of this knowledge is not confined to the Solar System, but extends into the stellar and interstellar environment, and amplifies our understanding of the processes which created the chemical elements, and the processes of evolution of stars and of the Galaxy.

Further Reading

"Meteorites: A Journey through Space and Time". Alex Bevan. Smithsonian Institute Press. 2002. For those without prior geological knowledge. Particularly good for explaining how radioactive dating works.

"Asteroids, Meteorites and Comets" by Linda Elkins-Tanton. Facts on File. 2006. Nice introduction to what can be learned and the links between these types of bodies.

"Moon Rocks and Minerals" by Alfred A. Levinson and S. Ross Taylor. Pergamon Press 1971. The Scientific results of the analysis of the rocks brought back from the Apollo 11 mission, with a preliminary look at samples from Apollo 12.

"Radiometric dating". Article in Wikepedia. Deals in useful detail with the processes which I have named when describing the results applied to lunar and meteoritic materials. https://en.wikipedia.org/wiki/Radiometric_dating#:~:text=Radiometric%20dating%2C%20radioactive%20dating%20or,incorporated%20when%20they%20were%20formed.

12

Comparing the Messages

Here you can compare directly the images of different classes of astronomical objects as observed via different methods, showing the complementary types of information that are offered which, when combined, can give us a deeper physical picture of each object. Most of the images are taken from the main text, but I have added extras to give a more complete picture for some of the objects.

12.1 Jupiter

As we have seen few images of Jupiter in the main text, I have used this appendix to give an amplified view of how we learn more by extending techniques. The images below include visible, radio, infrared, and ultraviolet observations, some from the ground, others from space, including one from the Juno fly-by which gave us our first clear look at the polar regions of the planet (Figs. 12.1, 12.2, 12.3 and 12.4). Jupiter is characterised by its extreme weather, due in large part to the mechanical energy input to its atmosphere from its very rapid rotation. This rotation, and the fact that its core is made up of hydrogen turned metallic by immense pressure, produces a powerful magnetic field, which gives rise to strong radio emission in its outer atmosphere.

© Springer Nature Switzerland AG 2021
J. E. Beckman, *Multimessenger Astronomy*, Astronomers' Universe,
https://doi.org/10.1007/978-3-030-68372-6_12

a

b

Fig. 12.1 (a) (Fig. 4.6b) Hubble Space Telescope image of Jupiter in the visible with a UV image of its polar aurora superposed around the polar region. The circular traces are caused by disturbances in the spiralling electrons caused by Jupiter's satellites Io, Ganymede, and Europa. The general features are atmospheric motions driven by the

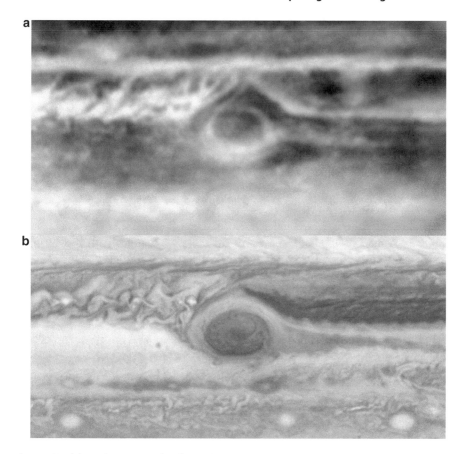

Fig. 12.2 (a) Radio image (top), made with the Karl Jansky Very Large Array. (b) Visible-light image (bottom) made with the Hubble Space Telescope, of Jupiter's famous Great Red Spot, a giant long-lived storm in the planet's atmosphere. The radio image shows the complex upwellings and downwellings of ammonia gas 30–90 kilometres below the visible clouds. Credit: de Pater, et al., NRAO/AUI/NSF; NASA

Fig. 12.1 (continued) rotation of Jupiter, nearly 13 km/s at its equator Credits: NASA/STcl. (b) This is a composite image made from JunoCam images taken during the 1st, 3rd, & 4th perijoves (closest passages to Jupiter) of NASA's Juno spacecraft. They were merged together using Photoshop and had the motion blurring reduced using a Wiener filter. This was the first time the polar regions of Jupiter have been observed and is made possible by the unique polar orbits of the Juno spacecraft. The imaging is performed within a programme where NASA provides the raw images and leaves it up to amateurs to perform the processing. The polar region is distinctly different from the familiar belts and zones of the temperate regions. The image shows a series of cyclones averaging 6000 miles in size. Credits: NASA/M.Sadler (Licensed under Creative Commons license 4.0). No modifications made to the original image

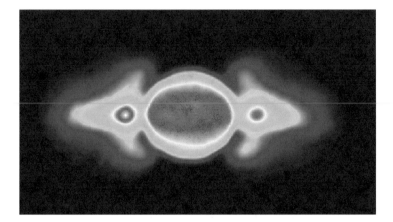

Fig. 12.3 A map of Jupiter taken at 13 cm radio wavelength with the Australia Telescope Compact Array. The central feature is thermal radiation from Jupiter's atmosphere, while the emission at both sides is synchrotron radiation emitted by rapidly moving electrons trapped in Jupiter's magnetic field. Credits: ATCA/G. Dulk et al

a   b

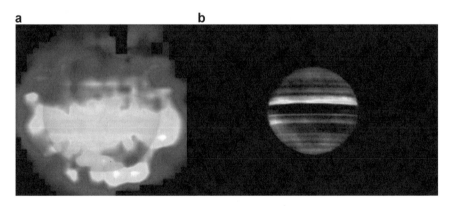

Fig. 12.4 (a) (Fig. 3.25) Left. The distribution of water vapour in Jupiter's outer atmosphere. The white zones are those with maximum water concentration, the cyan zones have rather less water, and the blue zones have least water. Notice that there is more water vapour over the southern hemisphere than over the northern hemisphere (see the text for the probable explanation). The map was made with Herschel's PACS instrument, at a wavelength of 66.4 μm Underlying is an image in visible light from the Hubble Space Telescope Credits: ESA/T. Cavalié et al. (University of Bordeaux/NASA/ Reta Beebe (New Mexico State Universit/Science Photo Libarary. (b) Right. Image of Jupiter at millimetre wavelengths taken by the ALMA interferometric telescope. To achieve the required brightness from the ground required several hours of integration, so the smaller features are blurred out by the rotation, leaving the major bands distinguishable. At these wavelengths the observations penetrate the upper ammonia clouds, and reach down to emission from water clouds some 90 km below. Bright yellow coding is for wamer material, darker coding is cooler material. Credits: de Pater et al. ALMA/ESO/NRAO/alma-jp

12.2   The Sun

As in all of astronomy, the study of the Sun was carried out in the visible waveband until recent decades. The complex information about temperature, composition, the presence and structure of the magnetic fields in the surface layer, the photosphere, could be obtained and modelled. The separate study of the oscillations of the Sun, giving information about its internal structure, was also performed in the visible. Limited information about the hot outer atmosphere, the chromosphere and the corona could be obtained during eclipses. The advent of ultraviolet and X-ray astronomy from space transformed out knowledge of the corona and the processes by which the Sun's mechanical energy is released into space via the solar wind. Studies of solar flares, and the discovery of coronal holes, and their relation with the solar wind were among the most important advances. In this appendix we have grouped the relevant information from the UV and X-ray imagers. The new messengers in the form of neutrinos have also given us information directly from the core of the Sun where its energy is generated by nuclear fusion reactions. Accumulating the detections over 20 years has produced an "image" of their distribution. The centre of this image does coincide with the centre of the Sun's disc, but as it is not possible to focus neutrinos, the response of the detection system can give only a very defocused picture, as we will see below (Figs. 12.5, 12.6, 12.7, 12.8, 12.9 and 12.10).

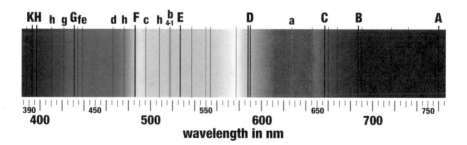

Fig. 12.5 (Fig. 1.9). The solar spectrum displayed visually showing some of its strongest lines. Fraunhofer's original simple naming, with capital letters A through K is shown. Astronomers still use the D for the strongest lines due to sodium, and H and K for singly ionized calcium. The wealth of information in these spectral lines is used to determine the composition, the local and global dynamics, the temperature and density structure of the layer which emits light to us, the photosphere. Spectra are at the heart of our physical understanding of most astronomical. Credit: Fraunhofer_lines.jpg: nl:Gebruiker:MaureenV Spectrum-sRGB.svg: Phrood~commonswiki Fraunhofer_lines_DE.svg: *Fraunhofer_lines.jpg: Saperaud 19:26, 5. Jul. 2005 derivative work: Cepheiden (talk) derivative work: Cepheiden, Public domain, via Wikimedia Commons

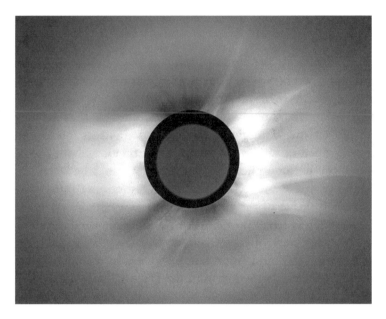

Fig. 12.6 (Fig. 3.6). Image of the Sun during the period of totality of the 1973 solar eclipse. Photo taken by Serge Kouchmy of the Institut d'Astrophysique, Paris. During the eclipse, whose totality as observed from the specially modified prototype Concorde supersonic airliner, 001, lasted 74 minutes, the corona and chromosphere were studied in the mid and far infrared with unprecedented angular resolution. This image does not give direct dynamical information, but there is a clear sense of outflow in the coronal plumes. Credit: Serge Koutchmy

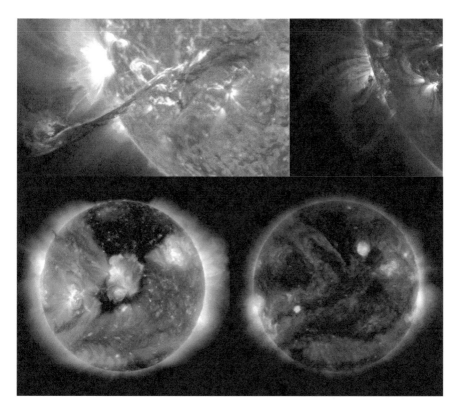

Fig. 12.7 (**a**) (Fig. 5.12) A solar flare on August 31st 2012 imaged in the Hα line from the Solar Dynamics Satellite Observatory which observed in the visible and ultraviolet range, into the softest X-ray region. Imaging from space, even in the visible, gives images of very high resolution which permit studies of flares in great detail. Credits: NASA/GSFC/ SDO (**b**) Figure 4.5a Zone of solar activity captured by NASA's solar dynamics observatory SDO (June 13th 2013) in the far UV at 17.1 nm wavelength The wealth of structures in the sun's upper atmosphere, the corona, is shaped by magnetic fields some of which loop out and back into the solar surface, the photosphere. You can sense the intense motions detected here, and seen in the videos obtained by the satellite. Where the magnetic lines do not loop down the particles are flowing outwards at high speeds to form an element of the solar wind. The local temperatures within the coronal loops can reach 10 million K. Credits: NASA/GSFC/SDO (**c**) (lower left) (Fig. 4.5b). The Sun from SDO with a large area coronal hole shown as dark in this far UV image, taken when the Sun was showing an intermediate level of activity. As a substantial part of it is facing the Earth a solar wind component was directed towards us on the day this was taken. Credits: NASA/GSFC/ SDO (**d**) (Fig. 5.10) The Sun. A composite of X-ray and ultraviolet images. The regions of greatest activity are in blue, as taken in 2015 by NASA's NuSTAR (Nuclear Spectroscopic Telescope Array) satellite, less active regions in green from JAXA's (the Japanese Space Agency's) HINODE satellite, and far ultraviolet emission (very close to the X-ray bands) from NASA's solar dynamics observatory, SDO Credits: JAXA/NASA

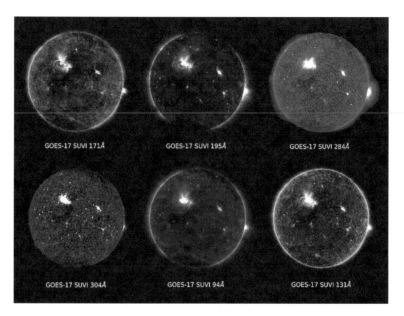

Fig. 12.8 (Fig. 5.11) A set of imagesof the Sun taken with NASA's GOES 17 through a series of filters between the far ultraviolet and the soft X-ray spectrum in which a flare is imaged in the top left quadrant of the solar disc. Each wavelength band samples a different range of heights in the corona Credits: NASA/WMO

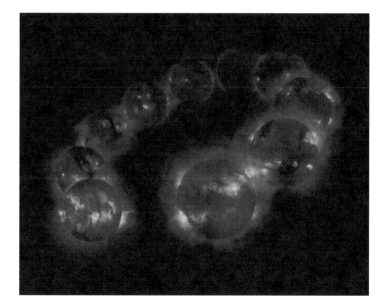

Fig. 12.9 (Fig. 5.13). X-ray images of the Sun during an 11 year sunspot cycle, starting in August 1991 at sunspot máximum taken by the YOKOH X-ray satellite. The more active regions, the flares, and the coronal holes are all present in these images, and we can see that at solar mínimum half way round the cycle the whole corona is much less brilliant in X-rays than at the two solar máxim, one at each end of the cycle. Credit: JAXA

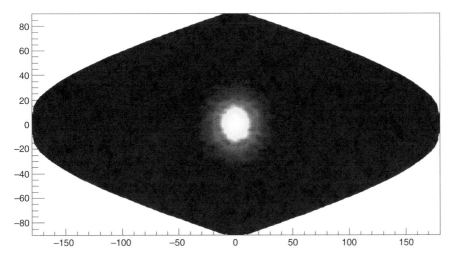

Fig. 12.10 (Fig 7.8). "Image" of the sun in neutrinos taken with the Superkamiokande neutrino telescope, during 5200 days of integration between 1996 and 2016. Scale in degrees. The large angular size is because the detection method cannot produce good angular resolution on the sky. The image diameter is some 40 degrees which is the effective angular response of the instrument. The colours represent neutrino counts, with respect to the position of the centre of the Sun, white high values, yellow intermediate values red low values, black background level. The visible Sun on this scale would be a small dot at the centre of the image. Credit: Kamioka Observatory, ICRR (Institute for Cosmic Ray Research), The University of Tokyo

12.3   The Milky Way

This section is composed of images of the Milky Way presented in different chapters of the book. Its aim is to show the differences revealed by using different wavelengths for the observations, explaining the importance of the fullest possible coverage. The Milky Way is the most complete example within the book of these comparisons. I have used a wide wavelength range, but most of the information is about the interstellar medium, rather than the stars. However the second image, which is essentially made up of millions of stars, is a recent result from ESA's GAIA astrometric satellite, whose main task is to map the positions and the distances of the stars in a major section of the Galaxy with unprecedented accuracy (Figs. 12.11, 12.12, 12.13, 12.14, 12.15, 12.16 and 12.17).

Fig. 12.11 (Fig. 3.2). The Milky Way photographed in visible light, with its plane making a projected arc over the dome of the 10.4 m Gran Telescopio Canarias, the largest optical/infrared telescope in the world. The centre of the Galaxy is just below the horizon, bottom left. Two notable features of this photo are the concentration of bright stellar light towards the central plane of the Galaxy, interrupted by the patchy darkness imposed by the interstellar dust which pervades the gas in the interstellar medium. This dust prevents us from detecting most of the stars in the Galaxy in visible light as we are ourselves very close to the plane. For this reason it is important to use other wavelengths to make adequate observations of the Milky Way as a whole. The curvilinear distance along the Milky Way plane in the image, from one extreme to the other is of order 15,000 light years. Credits: GTC/IAC/Daniel López

Fig. 12.12 Map of the Milky Way from observations by ESA's GAIA astrometric satellite. The number of stars sampled is 1.7 million, and GAIA's ability to measure distances enables astronomers to construct a 3d map from the data. But we need to remember that we are very close to the mid-plane of the galaxy, where the interstellar dust is most concentrated, which limits what we can see. The prominent dust lanes block our visión and reduce the range of our observations. The two smudges below the plane are the two Magellanic clouds Credits: Gaia Data Processing and Analysis Consortium (DPAC); A. Moitinho/A. F. Silva/M. Barros/C. Barata, University of Lisbon, Portugal; H. Savietto, Fork Research, Portugal

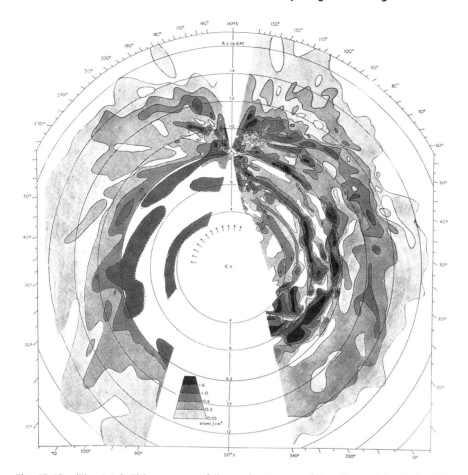

Fig. 12.13 (Fig. 1.4a). This was one of the earliest maps of the plane of the Milky Way made in 1958 using the newly exploited technique of mapping in atomic hydrogen with the emission line at 21 cm wavelength. The map is centred at the centre of the Galaxy, and the point of convergence in the upper part of the map is the position of the observer, i.e. the Earth. Two notable features of this map, firstly as atomic hydrogen is almost transparent the plane can be mapped to much greater distances than at visible wavelengths, and secondly the atomic hydrogen is widely dispersed between the stars, and allows continuous mapping. The denser concentrations of hydrogen are intuitively presented in darker shades. A peculiar feature of this method is that although data from radiotelescopes in both the northern and southern hemisphere were combined, there are conical areas towards the Galactic centre and anticentre where the map is blank. This is because the method of measuring distances to the hydrogen clouds relies on measurements of their velocities along the lines of sight, plus a model of the rotational velocity of the Galaxy (shown in Fig. 2.4b). Velocities along lines of sight tend to zero as we observe towards the centre or anticentre as the motions are essentially all circular, so no radial motion to measure. This basic drawback to good dynamical measurement of the Galaxy does not operate for external galaxies. The diameter of this map is some 80,000 light years. Credits: Oort. J. et al. 1958 MNRAS Vol 118 p.379

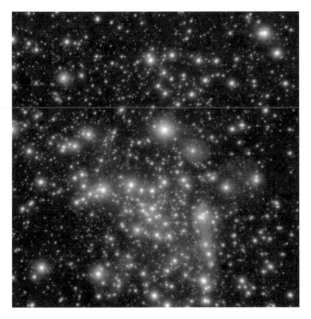

Fig. 12.14   (Fig. 3.4). Stars very close to the centre of the Galaxy They can be imaged using infrared techniques (so the colours in the image are not visible colours, but infrared colours coded into the visible for presentation) with adaptive optics for image sharpness. The infrared penetrates the dust and allows astronomers to detect and measure the motions of these stars around the central supermassive black hole. The sizes and velocities of the orbits, obtained over a period of 16 years, permit the calculation of the mass of the central object, some four million solar masses, and the limits to its size prove that it must be a black hole because the total mass of normal stars filling the volume would fall short of the measured mass by three orders of magnitude. This image is just over three light years across, a very different scale from the other images of the Milky Way presented here, a scale on which it is possible technically to achieve the ultra-high angular resolution needed to track the stars in their orbits. This work, together with precise kinematics of the orbiting stars, won the Nobel Prize in Physics, 2020 for Reinhard Genzel (shared with Andrea Ghez) Credits: Max Planck Institut für Extraterrestrische Physik (MPE)/Reinhard Genzel

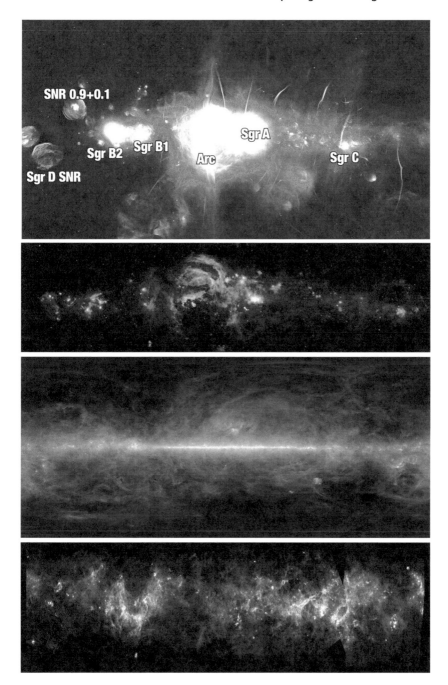

Fig. 12.15 (a) (Fig 2.21). The zone close to the plane of the Milky Way surrounding its centre, imaged at a radio wavelength of 23 cm by the MeerKat pathfinder telescope, a precursor to the great future Square Kilometre Array. The Galactic Centre is towards the constellation of Sagittarius, (abbreviation Sgr). The position of the supermassive

Fig. 12.15 (continued) black hole is close to the label Sgr A, and nearby radio sources are labelled Sgr B1, Sgr B2, and Sgr C. These are star forming clouds surrounded by molecular clouds. The more spherical objects, SgrD SNR and SNR 0.9 + 0.1 are supernova remnants (SNR): the results of expanding gas following supernova explosions. The arcs and filaments are not yet well understood. The horizontal size of the image is two degrees on the sky, four times the diameter of the moon. This corresponds to some 700 light years at the distance of the centre of the Galaxy. Credits: MeerKAT/ SAAO. (**b**) (Fig. 3.11). A composite infrared image of a zone in the plane of the Milky Way surrounding the Galactic Centre. It combines radiation observed with NASA's Spitzer Space Observatory, (represented in white) at 8 μm wavelength, radiation observed with the NASA/DLR SOFIA airborne observatory at 25 μm and 37 μm in blue and green respectively, and with ESA's Herschel Space Observatory at 70 μm (represented in red). The size of the image is comparable to that in Fig. 12.15(a), some 600 light years cross. The objects featured here are, however, mostly zones of intense star formation, in which the dust pushed into space from evolved stars is heated by the radiation from the younger stars and emits strongly in the infrared. Credits: NASA/ SOFIA/JPL-Caltech/ESA/Herschel. (**c**) (Fig. 3.13) Galactic cirrus. This image was taken by the first infrared astronomical satellite IRAS, and shows the plane of the Milky Way taken at 100 μm wavelength. At this wavelength the emission from the cool dust predominates. This cool dust in found mixed with neutral hydrogen gas, and you can see that this is concentrated towards the plane of the Galaxy. But where there is activity due either to zones of active star formation, or the explosive events assocated with dying stars, the gas and dust are pushed up to considerable heights above the plane, forming wispy "Galactic cirrus". The energy for the infrared glow also emanates from the active zones. The scale of the image is comparable to that in Fig. 12.15(a) and Fig. 12.15(b). Credits: NASA/SERC/NIVR. (**d**) (Fig. 3.24) Panoramic view of the zone around the plane of the Milky Way taken with ESA's Herschel satellite. The red represents the coolest densest interstellar dust, found especially in molecular clouds which will be the next sites of star formation. The other colours show warmer dust, heated by energy from young hot star clusters or the expanding explosive shells due to supernova explosions. The wavelength range covered is from far infrared: 55 μm to submillimetre: 672 μm. This image is nearly 2000 light years in the horizontal dimension. Credits: ESA/ PACS/SPIRE/S-Molinari, Hi-GAL project

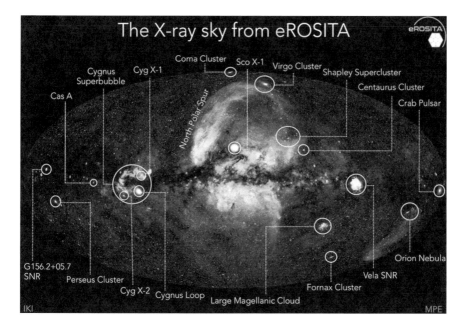

Fig. 12.16 (Fig. 5.23b). The Milky Way is the main feature of this map of the whole sky obtained by the Russian-German X-ray telescope eROSITA. False colours are used so that the lowest X-ray energies are in red, intermediate energies in green, and higher energies in blue. Individual freatures in the Galaxy such as the Cygnus Loop, the Vela supernova remnant, the Crab pulsar and the Orion Nebula star-forming region are shown, but there are also extragalactic objects, such as the Perseus cluster of galaxies, the Large Magellanic Cloud, the Coma and Virgo clusters of galaxies, and the Shapley supercluster of galaxies. The coordinates are chosen to place the plane of the Milky Way across the centre of the map. Distributed high energy X-rays come from gas where high energy electrons produced in violent events interact with magnetic fields. Credits Max Planck Institut für Extraterrestrische Physik: Jeremy Sander, Hermann Brunner and the eSASS team), IKI (Russia): Evgeny Churazov, Marat Gilfanov

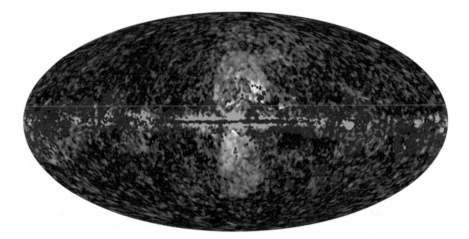

Fig. 12.17 Fermi bubbles and Galactic haze. This is a superposition of two all-sky maps at opposite ends of the electromagnetic spectrum. The blue areas represent the response of NASA's Fermi gamma-ray satellite at energies between 10 GeV and 100 Gev and features the "Fermi Bubbles" huge structures in the interstellar medium extending more than 30,000 light years above and below the Galactic centre (see also Fig. 5.23b), and a concentration of gamma-ray sources towards the Galactic Plane. The red areas show Planck's map of emission at radio frequencies of 30 and 44 GHz. We can see that this emission does overlap with the gamma-ray emission over major areas of the Fermi Bubbles, but is seen distributed more widely at relatively high galactic latitudes. Credits: ESA/Planck Collaboration (microwave); NASA/DOE/Fermi LAT/Dobler et al./Su et al. (gamma rays)

12.4   Supernova Remnants

12.4.1   The Crab Nebula

The Crab Nebula, within the constellation Taurus, is the first object in the famous catalogue of non-stellar objects produced in the mid eighteenth century by Charles Messier, so it is named M1. The supernova which produced it was seen and recorded by Chinese astronomers in 1054 as a "new" star which was visible in the sky for several months. It was rediscovered as a visible bright cloud of gas in the eighteenth century, and named the Crab a hundred years later by Lord Rosse (whose telescope features in Fig. 1.5) because of the overall shape it presents. It is one of the brightest radio sources in the sky, at first featured as Taurus A, and later as 3C144, its number in the 3rd Cambridge catalogue of radio sources. It is expanding so quickly that by taking images over a period of order 10 years the expansion can be measured directly. Using the emission from specific spectral lines, its velocity towards us can be measured by the Doppler effect. These two measurements give the distance, and can be used to "postdict" the time of the original explosion, giving a value of just over 1100 years ago, which agrees moderately well with the observed date of 1054. The prediction that the blue glow in the nebula is synchrotron radiation was made by Josef Shklovsky in 1953, and was shown to be correct when the radioastronomers measured its spectrum. When pulsars were discovered, and shown to be the neutron stars left after supernova explosions, Franco Pacini predicted that there should be a pulsar in the middle of the Crab Nebula, which was detected by the radioastronomers, and a few months later was the first pulsar to be detected in visible light. It is fairly nearby, in astronomical terms, 6500 light years away, and this has made it a test target for observations of intensity and polarisation over the full electromagnetic spectrum. In 2019 the highest energy gamma-rays ever detected and identified with a specific object were observed coming from the Crab nebula (Fig. 12.18).

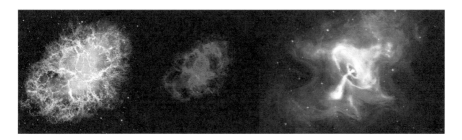

Fig. 12.18 **(a)** (Fig. 2.15b). A mosaic image of the Crab Nebula through three filters with the Hubble Space Telescope. The blue haze around the centre is synchrotron emission in the optical. The filaments of excited expanding gas show up in yellow from neutral oxygen, green from singly ionised sulphur, and red for doubly ionized oxygen. Credits: NASA/ESA/HST/J.Hester/A.Loll (ASU). **(b)** (Fig. 2.12). Radio continuum image of the Crab nebula supernova remnant taken with the Karl Jansky Very Large Array, in false colour. The emission highlights the filaments of rapidly expanding gas, which produce radio emission due to the spiralling of accelerated electrons in the magnetic field (synchrotron emission). This image shows the resemblance of the emission to the crab-like form from which its name derives. Credits: NRAO/AUI/NSF. **(c)** (Fig. 2.15a). This image is in combined optical emission by the Hubble Space Telescope, coloured purple, which shows the majority of the diffuse matter, infrared from the Spitzer Space telescope, coloured pink, which shows the dust and X-ray emission by the Chandra Space Telescope, coloured blue and white. The X-rays bring out the pulsar, the rapidly spinning neutron star in the centre of the nebula, and the effect of the rotating beam of radiation and particles on the surrounding gas in the nebula. Credits: NASA/ESA/HST/CXC/J. Hester et al.

12.4.2   The Veil Nebula

The Veil Nebula. A middle-aged supernova remnant. Even though it is not recent, using the Doppler method to measure its velocity towards us gives a value of 1500 km/sec for this expanding shell-like system. We can get an estimate of its size by measuring its distance, which gives a rough diameter of 90 light years. This allows us to calculate a rough time at which the expansion would have started, and this calculation gives 10,000 years ago. (always bearing in mind that these times are as we see them, and the events themselves occurred 2500 years previously because the nebula is 2500 light years away). Most of the light in the visible and ultraviolet comes from the densest parts of the nebula, where the expanding gas is being shocked by collision with the stationary gas which was in place before the explosion. However all the gas in the nebula is extremely hot, and radiates X-rays, as is shown by the difference between the third image below, and the upper two images (Fig. 12.19).

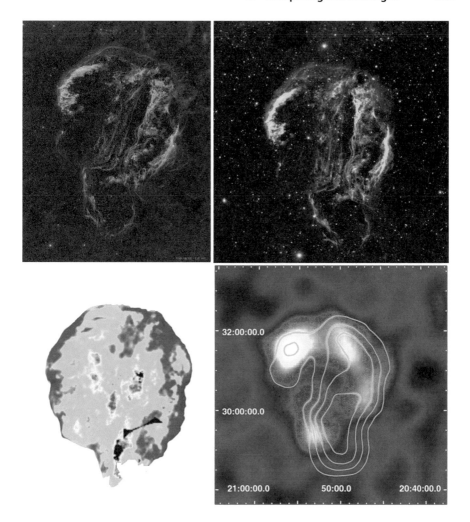

Fig. 12.19 (left to right, upper then lower) (**a**) (Fig. 1.18) The Veil Nebula, also called the Cygnus Loop, in visible light. The expanding remains of a supernova which exploded some 10,000 years ago. The diametral size is comparable to three times the lunar diameter. This image shows how a dedicated amateur can obtain beautiful data, especially over wide fields. This required 44 h of total exposure for an 8 panel mosaic, and combines light from two spectral lines: OIII in the visual and Hα in the red. Narrow band filters serve to minimise general background light. Taken from a backyard in Austin, Texas! Credits: Anis Abdul/astrobin.com. (**b**) (Fig. 4.8) Ultraviolet image of the Cygnus Loop, the expanding remnant of a supernova which occurred some 10,000 years ago, towards the constellation of Cygnus, the image diameter is over three times that of the moon. Taken with NASA's Galaxy Evolution Explorer UV camera. Credits: NASA/JPL-Caltech. (**c**) (Fig. 5.15). The Veil Nebula, also called the Cygnus Loop, a local volume of hot gas some 90 light years in diameter produced by a supernova which exploded 10,000 years ago, imaged in X-rays with the position sensitive proportional counter on ROSAT. Credits NASA/DLR/N. Levenson, S. Snowden. (**d**) FermiLAT gamma-ray image of the Veil Nebula, with the isophote contours of radio emission in green. Credits/H. Katagiri et al. 2011 ApJ Vol 741 p.44/Z. Tang et al. 2016, ApJ Suppl.Vol. 227, p. 82

12.4.3   Cassiopeia A

Cassiopeia A was one of the earliest radio sources to be found, by Martin Ryle and Francis Graham-Smith at Cambridge in 1948. It was detected in the visible wavelength range in 1950. Although it is one of the nearest supernovae to have exploded in the Milky Way during the last millenium there is no clear record of the explosion being observed, although there is a possibility that a star recorded at magnitude six in Cassiopeia around 1680 by the then Astronomer Royal, John Flamsteed, was in fact this supernova. It is possible that the star had ejected so much dust before it exploded that the visible light due to the explosion was largely absorbed. Observations at optical and radio wavelengths show that it is expanding very rapidly, at between 4000 and 6000 kilometres per second, and its gas has a temperature of 30 million degrees. It is not by any means the nearest observable supernova remnant, its distance is some 11,000 light years, but it is near enough to be spectacularly bright over the full range of the spectrum (Fig. 12.20).

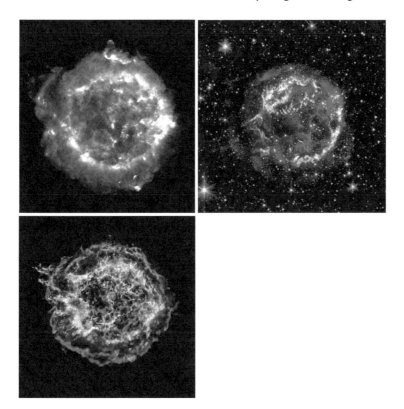

Fig. 12.20 (a) (Fig. 2.11) Image of the supernova remnant Cassiopeia A taken at 2.7 cm wavelength, (11 GHz frequency) with the Karl Jansky Very Large Array. The overall structure is clearly spherical, showing that the mean density of the gas surrounding the exploding star is isotropic, there are no major density variations in any given direction. But the considerable internal structure of the expanding cloud shows that on small scales the gas is quite inhomogeneous in density. Credits: *NRAO/AUI/NSF/L. Rudnick* et al. (**b**) (Fig. 3.17) Image of the Cassiopeia A supernova remnant taken by the Spitzer Space Telescope in the infrared. The image is a combination of exposures through three filters, at 3.6 µm, transformed to blue, at 4.5 µm, transformed to green, and at 8.0 µm, transformed to red. The orientation is a little different from that of CassA 1, but it is easily recognisable as the same object. We are seeing dust emission in this image, and the most intense dust emission comes from quite close to the most intense radio emission. This is because the radio continuum emission comes from collision-shocked gas, and the dust is compressed and heated by the shock fronts, emitting most strongly there. Credits: NASA/JPL (Whitney Clavin. (**c**) (Fig. 5.19a) Image of the Cassiopeia A Supernova remnant taken by the Chandra X-ray satellite. Colour coded X-ray bands: red, 0.3–1.2 keV,yellow 1.2–1.6 keV, cyan 1.6–2.26 keV navy 2.2–4.1 keV purple 4.4–6.1 keV. Again the orientation is a little different, but the object is recognisable. Hot, compressed shocked gas shines most strongly. The global structure is better revealed here, because the angular resolution is much better than in the VLA image, and somewhat better than in the Spitzer image. There is an outermost shock front with highest energy X-rays, in blue, showing where the initial rapidly expanding supernova gas has reached. The inner shocks are reflected, either on a large scale, (the broad diffuse purplish spherical shock) or on small scales (the tangled filaments), Combining physical information in all three images tells us a great deal about how supernovae affect large volumes of their surroundings. Credits: NASA/CXC/SAO

12.5   The Andromeda Galaxy

The Andromeda Galaxy, M31 in the classical catalogue of extended objects by Messier, is the largest galaxy in our local group, of which the Milky Way is the second largest. It is one of the few galaxies visible to the naked eye. Its relative proximity makes it a valuable subject for study. It shares most of the general characteristics of disc galaxies: a central spheroidal bulge, and a wide outer disc,with spiral arms marked by bright young stars and edged by lanes of dust in the interstellar gas. At optical wavelengths the main features of its structure are nicely revealed in Fig. 12.21. Disc galaxies are generally rotating about their central axes. A complete way to detect and measure their rotation is provided by radioastronomical measurements using the ubiquitous interstellar neutral atomic hydrogen. The observations of the hydrogen emission line give three parameters, one which measures the amount of hydrogen along the line of sight, the second which measures the velocity with respect to the observer, and the third which measures the internal turbulence of the hydrogen at each point of measurement. Figure 12.22a is a map of the first parameter, and shows that the hydrogen also takes a globally spiral form. Combining this information with that of the structure in Fig. 12.21 we can find the relation between the stars and the gas, which shows that the gas tends to pile up along the edges of the stellar arms, and that the denser the gas, the denser is the dust. In Fig. 12.22b the hydrogen is showing us the rotation of the galaxy, and as this gas is present over the whole disc, gives a more complete velocity map than we could obtain using light from the stars. Fig. 12.23, obtained in the far infrared by the Spitzer satellite, picks out the warm dust. This is found quite close to the stars, as this dust is heated by stellar radiation, and we can see that by comparing it with Fig. 12.24, in ultraviolet from the GALEX satellite, which picks out the hottest stars. There is a better general correspondence between the features in these two figures than between them and the features in Fig. 12.21. The dust in Fig. 12.21 is simply blocking the visible starlight, and so it includes both cold and warm components. The stars seen in Fig. 12.21 are a full range, from the youngest and hottest to the oldest and coolest. One of the tasks of a professional astrophysicist is to use photometric (and also spectroscopic) information to separate components and thus learn about the structural evolution of the galaxy.

Fig. 12.21   (Fig. 1.7) The Andromeda Galaxy, M31, our neighbour, in the visible.This an excellent overview, and includes the Andromeda Nebula's satellite galaxies M32 (close to the disc, left-centre) and M110 (lower right of centre) Credit: Ivan Bok, Wikipedia, Creative Commons

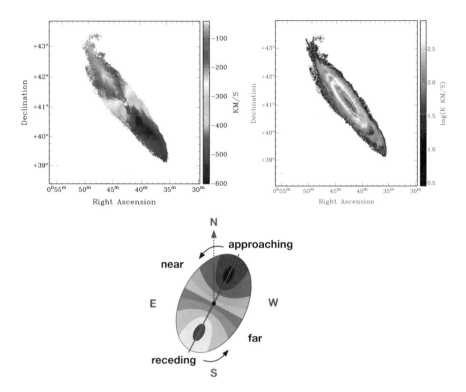

Fig. 12.22 (a) (Fig. 2.5a). Upper left. Map of the atomic hydrogen in our Local Group Galaxy M31, obtained using the 21 cm line, with the Dominion Astrophysical Observatory radio interferometer (Canada) plus the 26 m single dish. The intensity of the signal is coded in colour from yellow as the strongest to dark red as the weakest. The signal strength measures the integrated density of the column of atomic hydrogen observed at that point on the disc. (b) (Fig. 2.5b). Upper right. Map of the velocity of the atomic hydrogen in M31 from the data set used for the map in (a). The velocity observed is the component directed towards us, the observers, and is typical of what we observe for a rotating disc galaxy when the internal rotation in the plane of its disc is the dominant velocity. The colour coding is red for spin velocities directed away from us and blue fo spinr velocities directed towards us. Credits: C. Carignan et al. Astrophysical Journal Vol. 709. P. 139. 2009. (c) (Fig. 1.12b). Lower left. Schematic of the observed velocity field of an idealized rotating galaxy disc, inclined at an angle to the observer. The colours indicate the velocities of rotation relative to the centre of the disc, and range from blue for the largest measured velocity of approach to red for the largest measured velocity of recession. The schematic diagram is oriented differently from the observed images, but galaxies can present themselves at any angle to the line of sight. Credit Instituto de Astrofísica de Canarias-UC3

Fig. 12.23 (a) (Fig. 3.19) Upper. The Andromeda Galaxy, imaged by Spitzer at 24 μm wavelength. The spiral structure is very nicely shown because the warm dust detected at this wavelength is concentrated in the spiral arms where most star formation is occurring. We can also see clearly that the disc is warped, and distorted possibly due to

12.6 Epilogue

With these beautiful images showing the variety of appearances of astronomical objects, we have arrived at the end of the guided tour of our gallery of the Universe. Just as our appreciation of local reality, which we sample with our sense of sight, is augmented by our senses of hearing, smell, taste, and touch, astronomy in the twenty-first century is able to use a wide range of new "senses", to receive the multiple messages which let us probe our cosmic environment and enhance our understanding of its complexity.

Fig. 12.23 (continued) interactions with its satellites. This figure has been rotated to fit the orientation of the UV image below. Credits: NASA/JPL. (**b**) (Fig. 4.10). Lower. Ultraviolet image of the Andromeda Galaxy from the GALEX satellite The global morphology in the UV is quite similar to that in the IR because the UV shows the presence of the hottest most massive stars, which heat the dust in their surroundings so that it emits in the IR Credit NASA/JPL Caltech

Appendix: Astronomical Distance and Energy Units

Astronomical Distances

The fundamental distance unit in astronomy is the light year, the distance light travels in one year at its speed in vacuum of 299,792.458 km/sec (normally written in physics as km s^{-1}). For the purpose of this book we can use the value of 300,000 km s^{-1} as a good approximation. The year which is used to define a light year is the "Julian year" which is 365.25 days, each of 86,400 seconds. This makes a total of 31,557,600 seconds. With these definitions one light year is 9,460,730,472,580.8 km (which is 9.461 Petametres).

Because of the history of making measurements to stars, the unit of distance most commonly used by astronomers until recently is the parsec, which has a value of 3.26156 light years. The method on which all astronomical distances are based is that of triangulation, looking at an object from two positions, at a known separation, and measuring the difference in the angle at which the object is observed. In astronomy the separation of the two positions is the mean Sun-Earth distance, and an object which appears to move on the sky by one arc second when observed from points with this separation is defined to be at a distance of 1 parsec. A parsec is, as we see, bigger than a light year, so it was used for the enormous distances to astronomical objects, although as physics plays an ever more important role in astronomy, there has been a recent tendency to quote in light years.

Abbreviation for parsec is pc, with corresponding kiloparsec kpc, Megaparsec Mpc, Gigaparsec, Gpc etc.

© Springer Nature Switzerland AG 2021
J. E. Beckman, *Multimessenger Astronomy*, Astronomers' Universe,
https://doi.org/10.1007/978-3-030-68372-6

To give some relevant examples:

Distance to the nearest star, Proxima Centauri 4.243 light years = 1.3 pc.
Distance to the centre of the Galaxy: just over 26 thousand light years, i.e. close to 8 kpc.
Distance to the Large Magellanic Cloud 158 thousand light years, 48.4 kpc.
Distance to the Andromeda galaxy (within our local group) 2540 million light years 780 kpc.
Distance to the centre of the nearest big galaxy cluster (Virgo cluster) about 65 million light years 20 Mpc.
Distance to the furthest currently known galaxy 13.3 thousand million light years 4 Gpc.

Energies of Particles and Radiation at X-Ray and Gamma-Ray Wavelengths

The unit of energy is the electron-volt, written eV. Its value is: 1.602×10^{-19} joules. It is the energy gained by an electron when its potential increases by 1 volt. An energy of 1 electron-volt (eV) corresponds to a near infrared photon of wavelength 1240 nanometres (nm) and a photon of visible light in the red, at 620 nm has an energy of 2 eV.

Multiples of electron volts:

Name	kiloelectron-volt	Megaelectron-volt	Gigaelectron-volt	Teraelectron-volt
Abbreviation	keV	MeV	GeV	TeV
Multiple of eV	1000 (10^3)	1,000,000 (10^6)	1000 million (10^9)	10^{12}

Name	Petaelectron-volt	Exaelectron-volt
Abbreviation	PeV	EeV
Multiple of eV	10^{15}	10^{18}

X-rays have energies in the kiloelectron-volt (keV) range.
Gamma-rays have energies in the Megaelectron-volt (MeV) range and higher.

Index of Objects

© Springer Nature Switzerland AG 2021
J. E. Beckman, *Multimessenger Astronomy*, Astronomers' Universe,
https://doi.org/10.1007/978-3-030-68372-6

Name Index [1]

[1] Names in bold face are of Nobel Laureates

© Springer Nature Switzerland AG 2021
J. E. Beckman, *Multimessenger Astronomy*, Astronomers' Universe,
https://doi.org/10.1007/978-3-030-68372-6

Subject Index

© Springer Nature Switzerland AG 2021
J. E. Beckman, *Multimessenger Astronomy*, Astronomers' Universe,
https://doi.org/10.1007/978-3-030-68372-6